MITSUBISHI 2.0 LITER 4-CYLINDER ENGINE

TABLE OF CONTENTS

 ISBN 0-936168-53-6

NOTES

MITSUBISHI 2.0 LITER 4-CYLINDER ENGINE

FOREWORD

This manual will introduce interested service technicians to the Mitsubishi G63B 2.0 Liter engine and serve as a companion training aid to the service manual.

Accordingly, the content of this manual is essentially descriptive with special attention given to those areas which differ from the 1982 G52B 2.0 Liter engine which the 1983 G63B engine is replacing in the RAM-50 truck. Servicing sequences in a highly condensed form are included to further highlight those areas where servicing requirements and procedures are different for the new engine.

Special Tool requirements, adjustment and tightening specifications plus a comprehensive listing of engine component specifications are also included for references purposes of the reader.

While the information in this manual was correct at the time of publication, certain items or specifications may change as a result of continuing product improvement. It is suggested the reader keep this manual current by noting these product changes on the blank ruled pages provided after the final section.

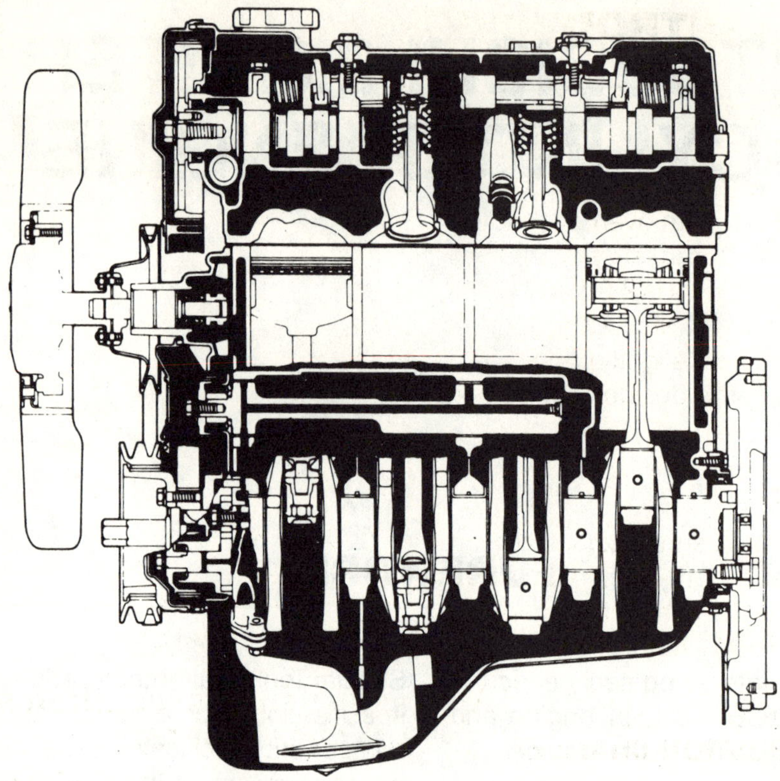

1983 Mitsubishi 2.0 Liter Engine (Side View)

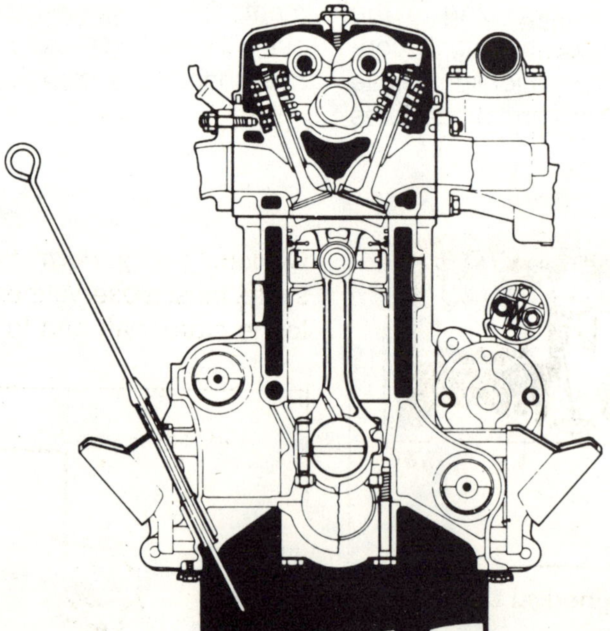

1983 Mitsubishi 2.0 Liter Engine (Front View)

MITSUBISHI 2.0 LITER 4-CYLINDER ENGINE

SECTION I — INTRODUCTION

The 1983 Mitsubishi 2.0 Liter engine is a gasoline-fueled, carbureted in-line Four, with overhead valves and a forged, counterweighted crankshaft supported in five main bearings. Displacement is 2.0 liters (1,997 Cm^3), essentially the same as 1982 2.0 Liter (although its bore and stroke differs slightly) and it continues to offer many of the same design advantages of the '82 engine such as:

OVERHEAD CAMSHAFT

Simpler in configuration than pushrod-actuated overhead valves, the overhead cam-overhead valve system is easier to service. Its lighter reciprocating mass also permits higher RPM's before the onset of valve float.

HEMISPHERICAL COMBUSTION CHAMBER

The hemispherical chamber permits the use of opposed valves with large valve head diameters affording a very generous cross-chamber flow pattern for excellent breathing. Also, since the hemisphere encloses a combustion chamber with minimum surface area, heat loss through chamber wall is also minimized, boosting thermal efficiency.

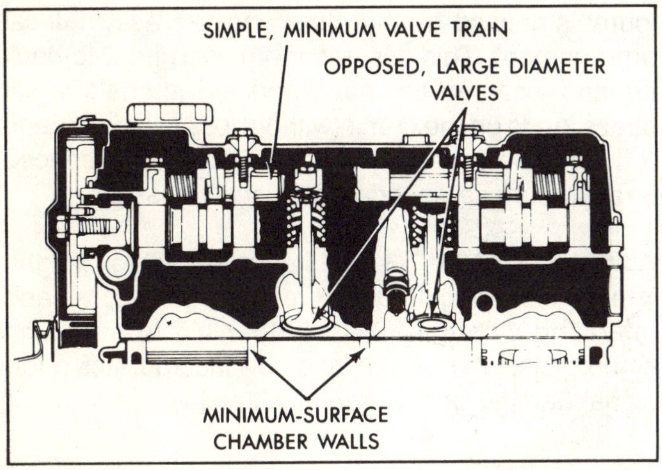

Efficient Overhead Cam, Hemispherical Combustion Chamber Head Design

JET VALVE

Timed to open with each cylinder intake valve, this small valve induces a swirling mixture flow that improves combustion efficiency and helps scavange incompletely burned fuel products.

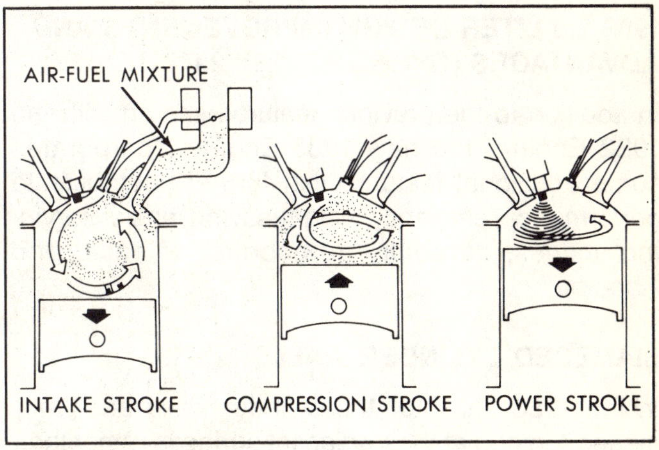

Jet Valve Air Flow

SILENT SHAFTS (COUNTER-ROTATING BALANCE SHAFTS)

These shafts, pioneered by Mitsubishi, help overcome an inherent shortcoming of the in-line four-cylinder engine: the inability to cancel out reciprocating vibration forces (from connecting rods) with any form of conventional crankshaft design or crankshaft counterweight placement. Because the two balance shafts of this Mitsubishi 2.0 Liter rotate at *twice* crankshaft speed and in *opposite* rotation they can help neutralize *radial* forces (up-and-down, side-to-side, etc.) and *torsional* forces (those that rock the engine around the crankshaft).

Obviously, to perform their function properly, both shafts must preserve their design timing with respect to the crankshaft and to each other.

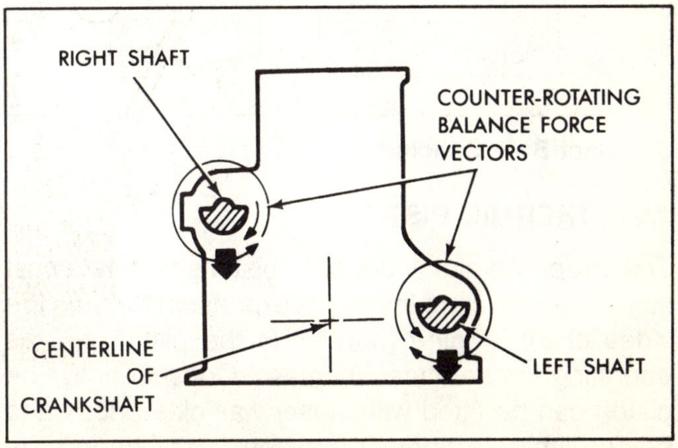

Balance Shafts in Operation

SECTION I — INTRODUCTION

1983 2.0 LITER DESIGN IMPROVEMENTS AND ADVANTAGES (OVER THE 1982 2.0 LITER)

In addition to the previous features shared with the 1982 Engine, the new 1983 Engine incorporates some additional features that further contribute to performance, efficiency, weight-saving, easier servicing access, durability and control of noise and vibration.

SIAMESED CYLINDER WALLS

In the 1983 2.0 Liter the cylinder walls are ("siamesed") joined at their adjacent surfaces rather than separated. This design, used so successfully in Chrysler's 1.7- and domestic 2.2 Liter engines, permits a shorter and lighter-weight engine. It also turns the cylinder bank into a single, unitized strength structure that enhances engine rigidity. The slight out-of-round wall distortion which may result from uneven cooling is easily accomodated by the new autothermic pistons.

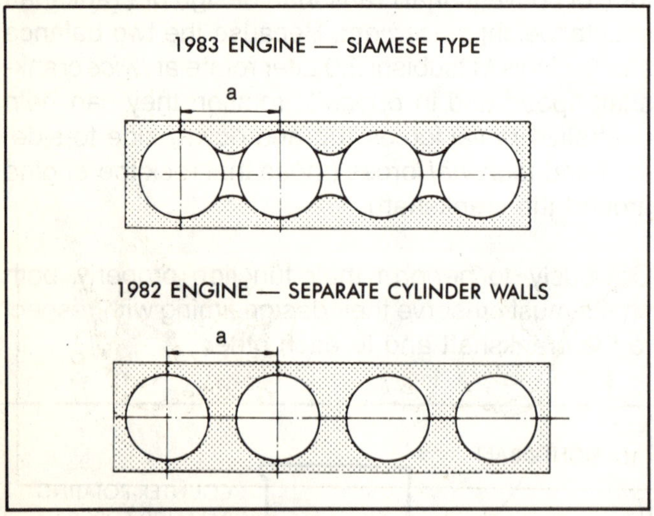

Compact Bore Spacing

AUTOTHERMIC PISTONS

The three-ring aluminum-alloy pistons feature a cast steel invar structure that limits expansion towards the sides of the engine (parallel to the piston pin) as operating temperatures increase. Consequently, the piston can be fitted with closer wall clearances and helps eliminate annoying piston "slap" during the starting and warm-up period.

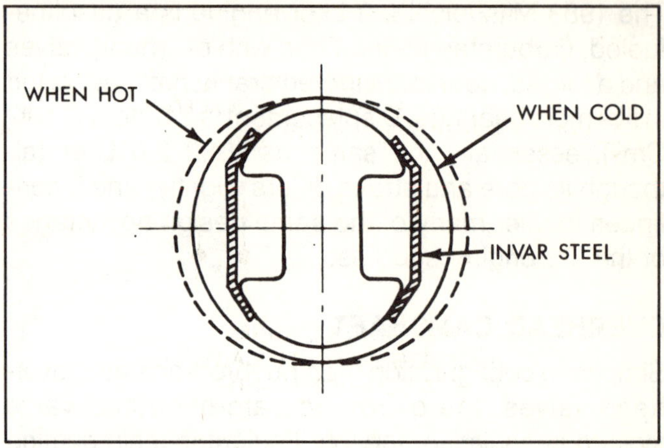

Autothermic Piston

EIGHT-COUNTERWEIGHT CRANKSHAFT

On the '83 2.0 Liter Engine, the crank incorporates a counterweight in each cheek aside each rod journal giving a total of eight counterweights. In the four-counterweight '82 engine crank, only *one* cheek in the pair around each rod journal has a counterweight (and it is larger). So, while both cranks are *statically* balanced, only the 1983 2.0 Liter is *dynamically* balanced. Its symetrically-located counterweights virtually eliminate the crank flexing between main journals caused by the older engine's assymetrical arrangement. This assures more even front-to-back bearing wear on the main and rod journals radial stress loads on the crank (without loss of rigidity) and reduce main and rod journal diameters for reduced bearing surface speeds and friction losses.

Advantages of the '83 2.0 Liter crank: lighter weight, improved bearing durability, reduced friction loss and operating vibration. As a further aid to operating smoothness, the new engine now incorporates a torsional damper in the crankshaft pulley.

SECTION I — INTRODUCTION

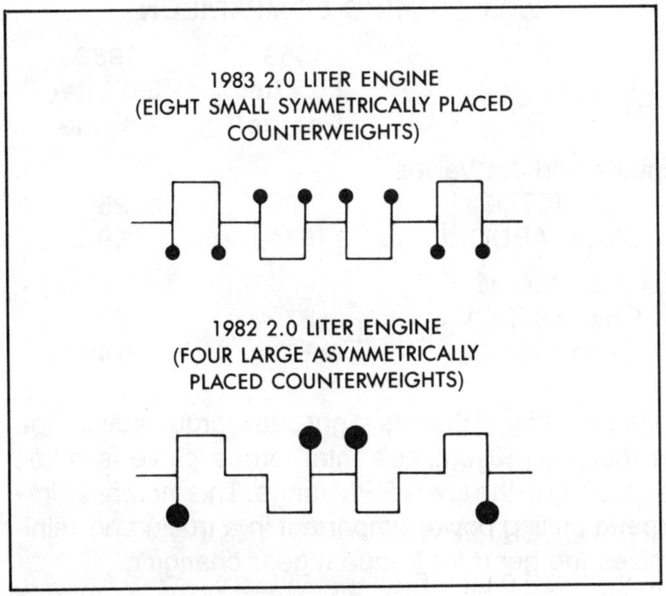

Dynamically Balanced Crankshaft

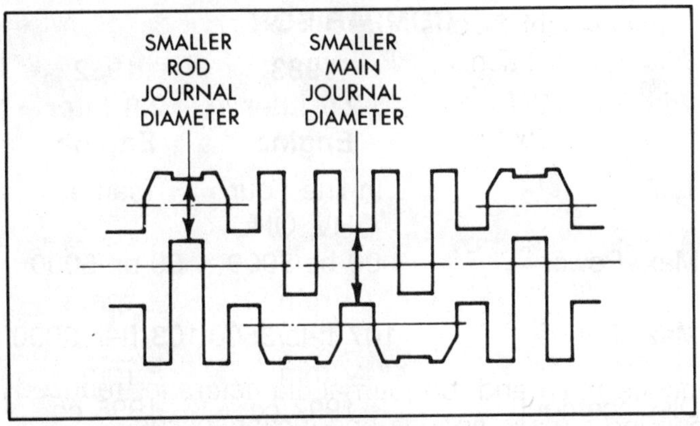

Reduced Bearing Friction Loss

SHORTER, LIGHTER CONNECTING RODS

With the smaller crankshaft counterweights, it was possible to raise the crankshaft axis closer to the cylinder walls, permitting the use of lighter and shorter connecting rods. This, in turn, permitted a reduction in overall engine height (and further reduction in weight). The lighter weight of the rods, in conjunction with the lighter crank, also affords a slight increase in performance since there is less rotating mass to absorb acceleration horsepower.

TIMING BELTS REPLACE TIMING CHAINS

One of the most evident changes with the '83 2.0 Liter Engine is the use of timing *belts* to drive the camshaft, oil pump and balance shafts. To facilitate this change, the balance shafts have been relocated in the block and the oil pump moved from the lower right to the lower left of the front end of the block. Accordingly, the driven gear of the oil pump now drives the *left* rather than the right balance shaft so it is on the *left* shaft which rotates in the direction opposite to the crankshaft. Technicians familiar with the Mitsubishi 1.6-liter Silent Shaft engine will recognize the strong similarity of its belt drive layout to that of the '83 2.0 Liter and how it differs from the chain drive layout of the '82 engine.

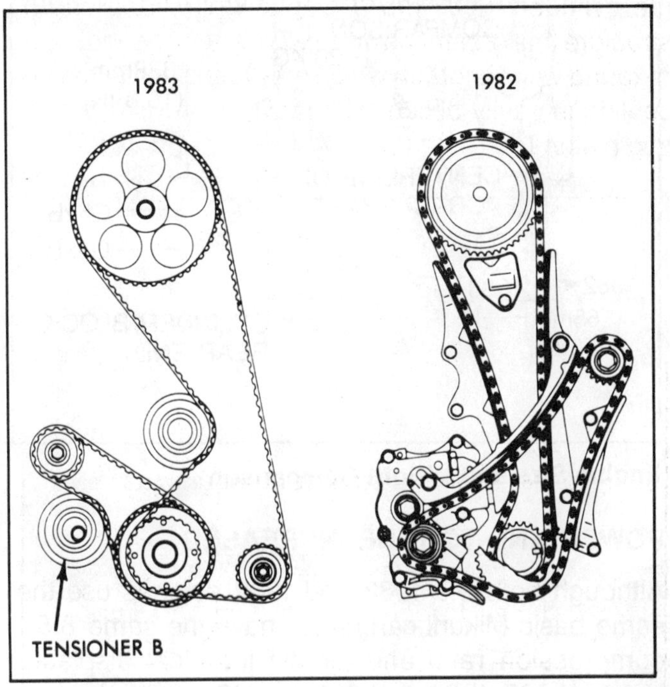

Timing Belts Replace Timing Chains

The advantages of the belt drive are that it eliminates a major source of engine noise (the timing chains), requires no engine oil lubrication and is a more simple and lighter-weight arrangement.

As a further aid to servicing ease, the on-vehicle adjustment of the front timing belt (which drives the camshaft and oil pump) can be adjusted without removing the lower front cover for the timing belt.

SECTION I — INTRODUCTION

REDUCED ENGINE WEIGHT AND SIZE ADVANTAGES

The 44-pound (20 kg) weight reduction offers owner benefits (increased payload, or slightly improved performance, demands on tires and brakes). The *size* reduction benefits service technicians by providing more working clearance around the engine for servicing.

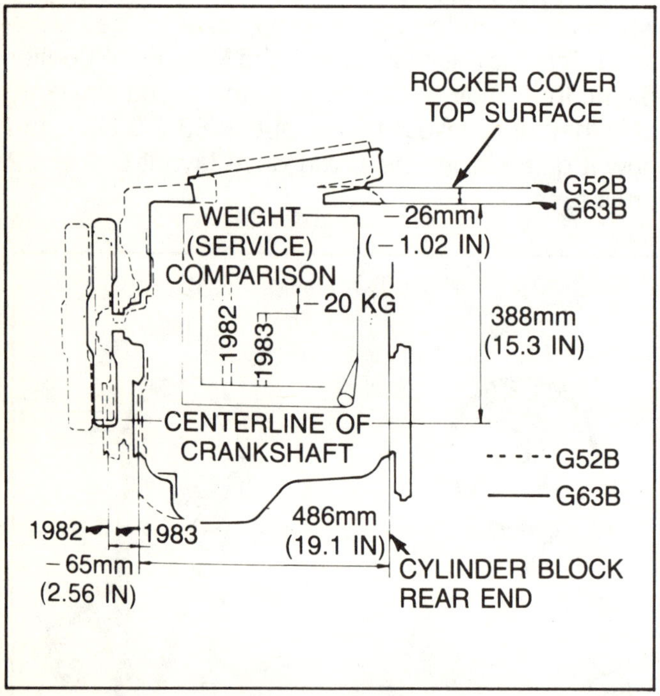

Engine Size and Weight Comparison

POWER AND TORQUE INCREASED

Although both the 1983 and 1982 engines use the same basic Mikuni carburetor, have the same 8.5:1 compression ratio and almost identical displacements (1,997, 1983 vs. 1,995) the '83 engine shows a slight increase in torque (almost 4%). The torque increase seems to result from valve timing characterized by reduced valve open and valve overlap times which provides a more efficient match with the induction and exhaust systems.

VALVE TIMING COMPARISON

	1983 2.0 Liter Engine	1982 2.0 Liter Engine
Intake and Jet Valves		
Open (BTDC)	19°	25°
Close (ABDC)	57°	59°
Exhaust Valves		
Open (BTDC)	57°	70°
Close (ABDC)	19°	14°

More significant than its slight peak torque advantage is that the '83 engine's total torque curve is fatter, expecially in the lower RPM range. This increases low speed pulling power (important in a truck) and minimizes the need for frequent gear changing.

GENERAL ENGINE SPECIFICATION COMPARISON

	1983 2.0 Liter Engine	1982 2.0 Liter Engine
Type	In-line Four, OHV, OHC	(same)
Max. Power	90 hp/5000 rpm	90 hp/5000 rpm
Max. Torque	107 ft-lb/3500 rpm	103 ft-lb/3000 rpm
Displacement	1997 cc (121.8 ci)	1995 cc (121.7 ci)
Bore	85 mm (3.35 in.)	84 mm (3.31 in.)
Stroke	88 mm (3.46 in.)	90 mm (3.54 in.)
Compression Radio	8.5:One	(same)
Carburetor	Two-bbl	(same)
Firing Order	1-3-4-2	(same)
Dressed Engine Weight (dry) with man. trans.	140 kg. (308 lbs.)	160 kg. (352 lbs.)

SECTION II — SPECIAL TOOLS

In addition to the usual metric-based tools, certain additional Special Tools are required for routine adjustment and repair servicing of the 1983 2.0 Liter engine. Most of these Special Tools are already in shop tool inventories for servicing current Mitsubishi engines. However, there are six *additional* Special Tools needed for servicing. These are:

- Cylinder Head Bolt Wrench (MD-998051)
  (NOTE: This tools also fits the 2.3 Liter Diesel engine head bolts)

- Silent Shaft Bearing Puller (MD-998371)

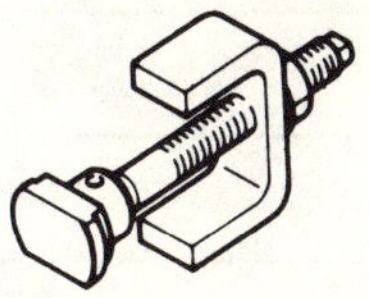

- Guide Plate (MD-998374)
  This tool is used with the following two Silent Shaft tools)

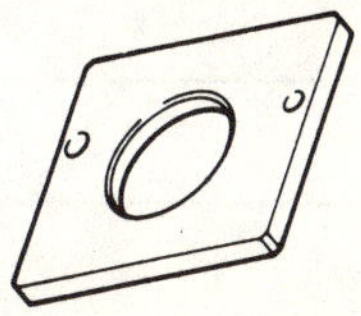

- Silent Shaft Bearing Puller (MD-998372)

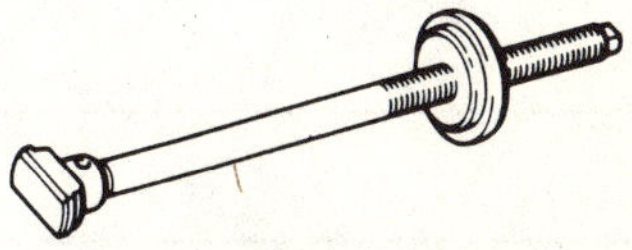

- Silent Shaft Bearing Installer (MD-998373)

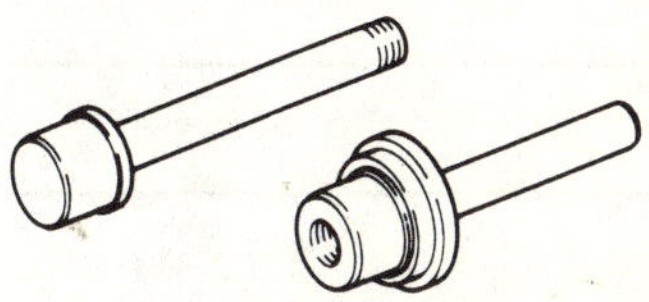

- Crankshaft Rear Oil Seal Installer (MD-998376)
  (This tool is used with C-4291 Handle)

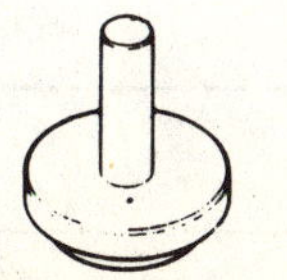

NOTES

SECTION III — CYLINDER HEAD — CYLINDER BLOCK ASSEMBLY

The '83 2.0 Liter head and block assembly is substantially different from that of the '82 engine. The three major changes are:

- Timing belt drive (rather than timing chain) for the camshaft and the newly-relocated silent shafts and oil pump (see "a").
- The use of siamesed (rather than separate) cylinder walls (see "b").
- The use of a full eight-counterweight (rather than four) crankshaft (see "c").

The result with these and other engineering improvements is that the 1983 2.0 Liter engine is:

- Lighter
- More compact
- More powerful and efficient
- Quieter
- Smoother
- More durable . . .

. . . than the 1982 engine it replaces.

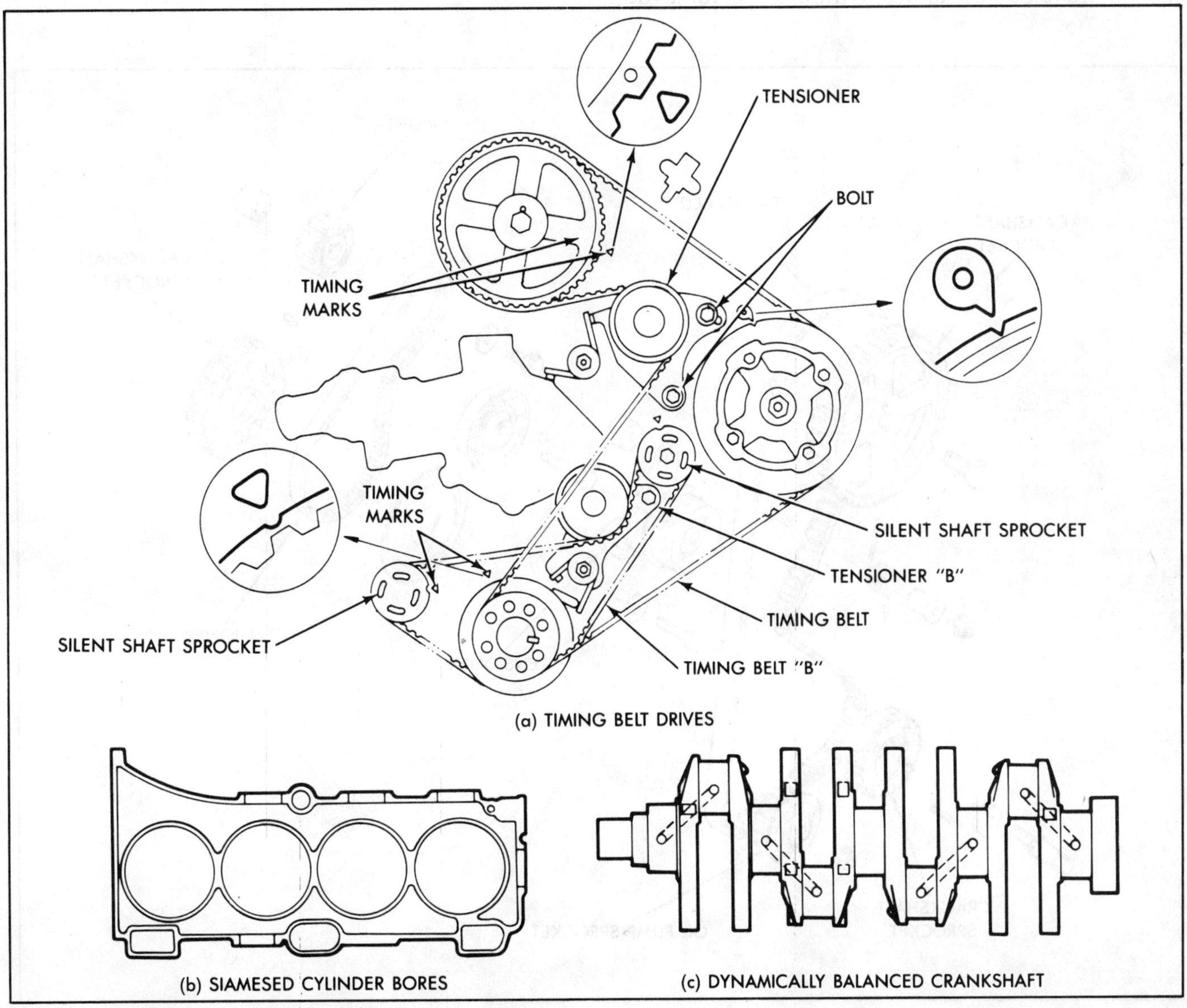

The Three Major Head/Block Changes for 1983

SECTION III — CYLINDER HEAD — CYLINDER BLOCK ASSEMBLY

TIMING BELT DRIVE SYSTEM

The 1983 engine's belt drive is simpler and easier to service with fewer parts than the chain drive of the '82 engine. Another major belt drive servicing advantage is that it is self-lubricating rather than running in an oil bath required for the chain drive. Because the drive covers (not shown) need only be dust-tight rather than oil tight, it eliminates the bothersome trimming or replacement of gaskets that is required with oil-tight covers.

Since the system is dry rather than oil-coated it's much more pleasant to work with. Belts removed during disassembly are dry and clean so they can be inspected more quickly. Belt-type sprockets are not as subject to excessive wear or missing teeth. Reassembly and readjustment are also simpler and quicker.

BELT DRIVE LAYOUT

The timing (front) belt drives the camshaft sprocket and the oil pump sprocket (which in turn drives the left silent shaft). The "B" (rear) belt drives only the right silent shaft. Tensioners are provided for both belts.

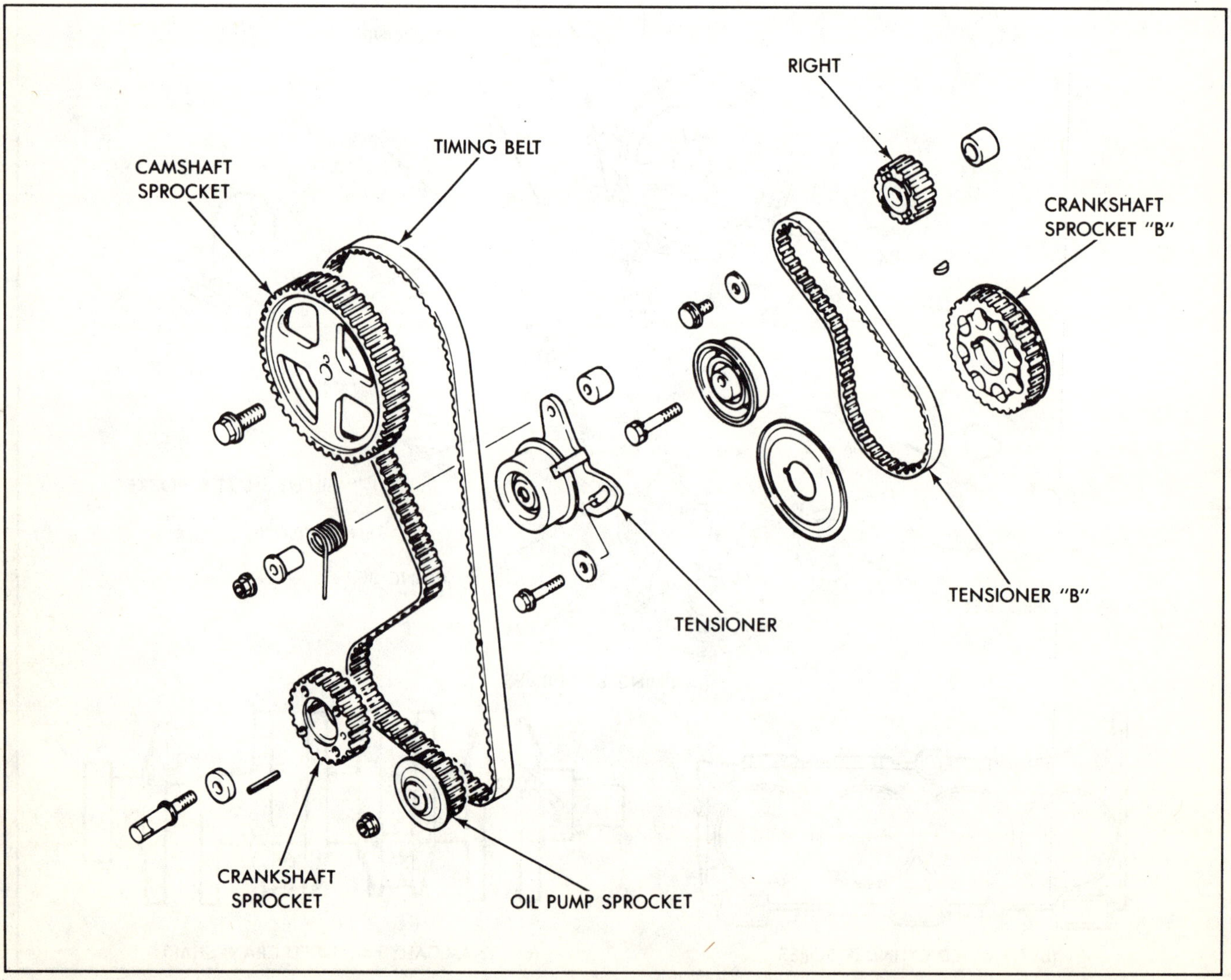

Timing Belt Train

SECTION III — CYLINDER HEAD — CYLINDER BLOCK ASSEMBLY

INSPECTION

Both belts can be inspected on-vehicle by removing the timing belt front upper cover and rotating the crankshaft until the full length of each belt has come into view. Belts should be replaced if there is visual evidence of:

- Aging or hardening of back surface
- Cracking or separation
- Abnormal wear on tooth surfaces or belt sides
- Missing teeth.

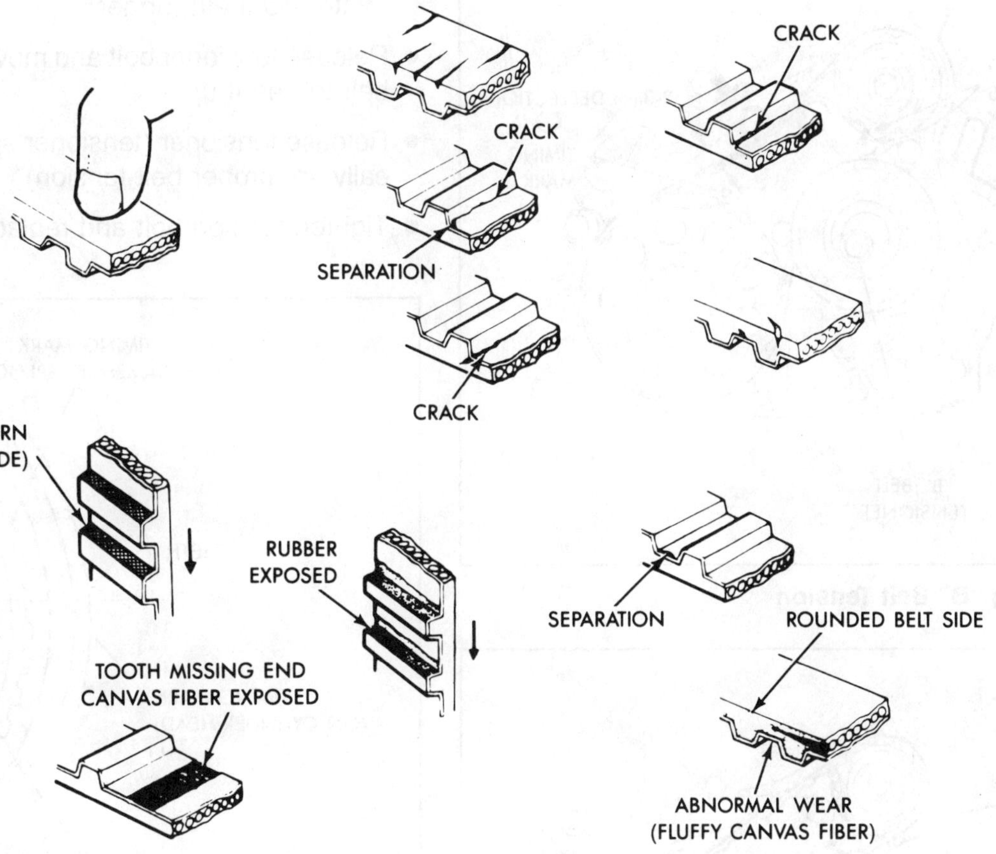

Inspection

SECTION III — CYLINDER HEAD — CYLINDER BLOCK ASSEMBLY

"B" (REAR) BELT TENSION AND READJUSTMENT

During inspection, the "B" belt should show about 2 to 3 mm of deflection. Tension can be readjusted by loosening the eccentric bolt on the tensioner rotating the eccentric until tension is restored, retightening the bolt. This adjustment can be performed only with the timing belt *lower* cover removed which requires that the engine be removed from the truck.

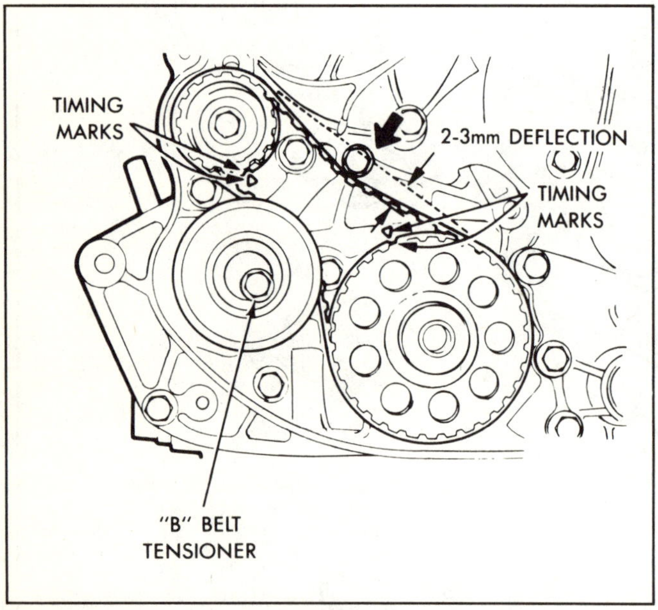

Checking "B" Belt Tension

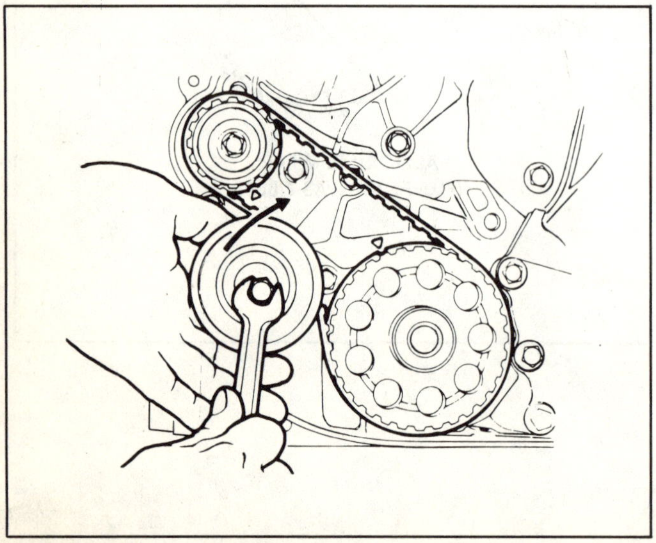

Adjusting "B" Belt Tension

TIMING (FRONT) BELT TENSION AND READJUSTMENT

Because the timing belt is longer and handles a heavier load, it may require more frequent tension inspection and adjustment. A simplified on-vehicle procedure for adjusting timing belt tension is provided:

- Remove front upper cover and two access covers in front lower cover
- Index timing marks on camshaft sprocket, then rotate *two* teeth further.
- Release tensioner bolt and move tensioner *toward* belt to free it up
- Release tensioner (tensioner spring will automatically set proper belt tension)
- Tighten tension bolt and replace covers

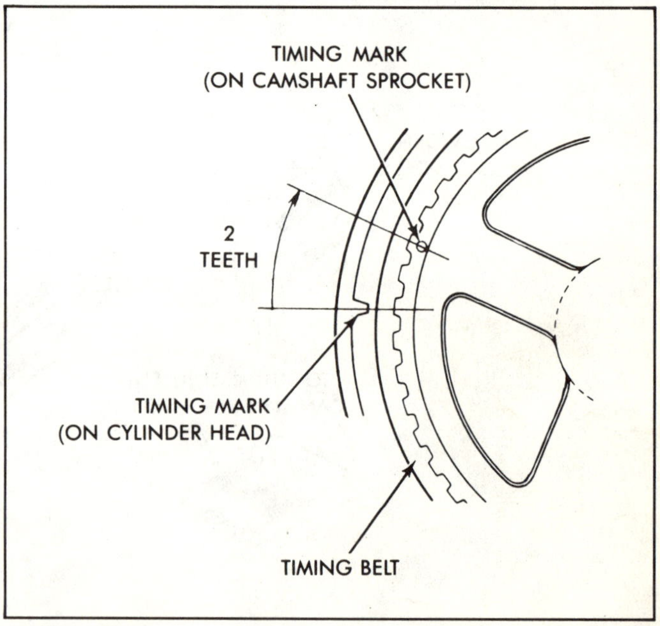

Timing Marks

REMOVAL OF BELTS, SPROCKETS AND TENSIONERS

These operations require both front covers be removed from the engine. Since the front lower cover cannot be removed on-vehicle, the engine must be removed.

SECTION III — CYLINDER HEAD — CYLINDER BLOCK ASSEMBLY

FRONT CASE, OIL PUMP AND SILENT SHAFTS

The *front case* serves as the front cover for the engine block and the foundation for the belt drive covers and tensioners, oil filter pad and relief valve assembly. It also accommodates the front cover of the oil pump and the cover for the front bearing of the right silent shaft.

The gear-type *oil pump* is mounted on the left rear surface of the front case. The oil pump's *driven* gear drives the left silent shaft and its bearing also serves as the shaft's front bearing.

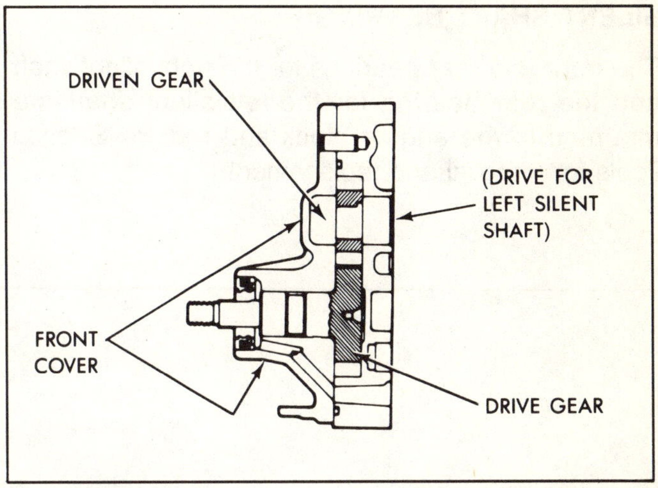

Oil Pump

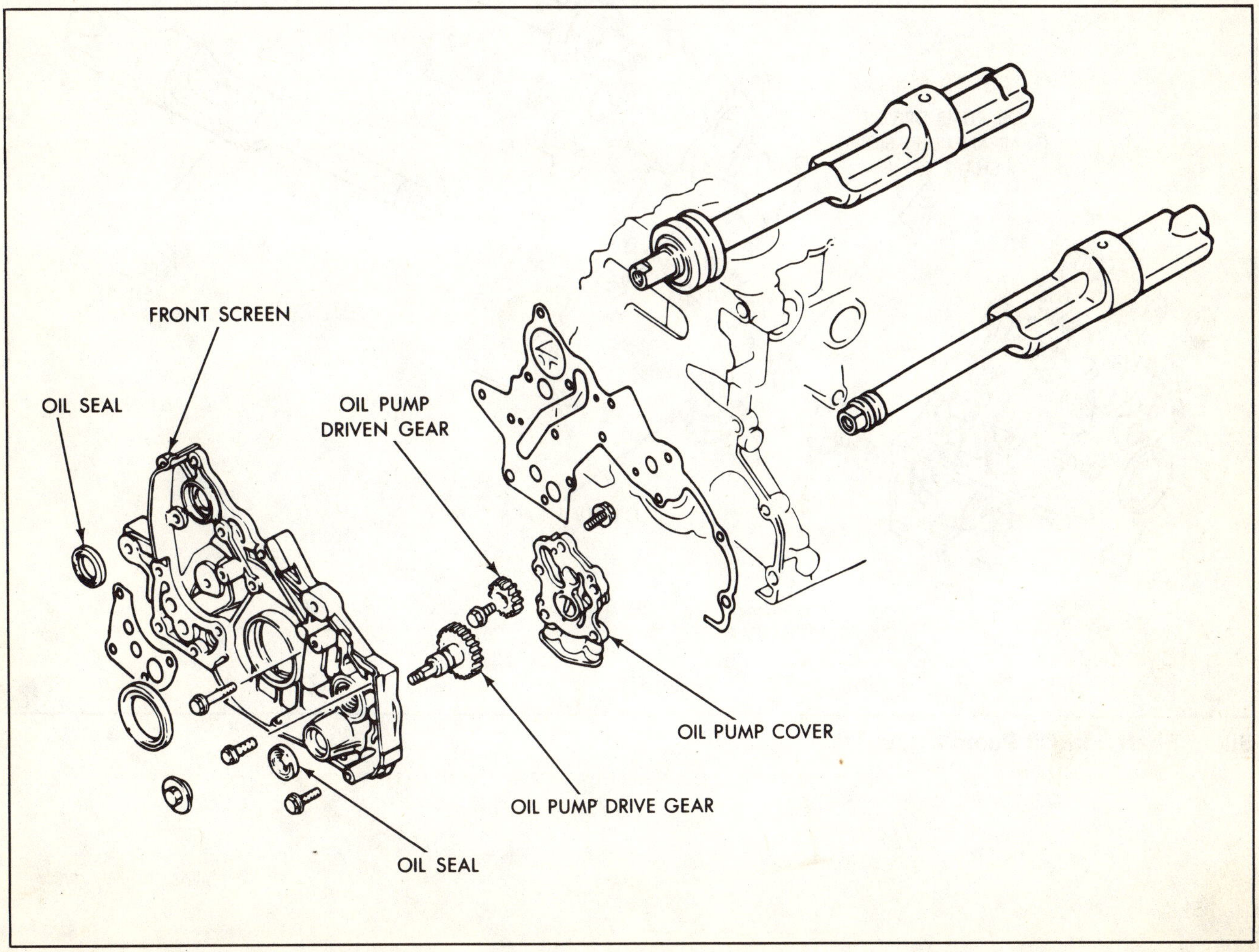

Front Cover and Related Components

SECTION III — CYLINDER HEAD — CYLINDER BLOCK ASSEMBLY

SILENT SHAFT BEARINGS

The front and rear bearings for the right silent shaft and the rear bearing for the left silent shaft are mounted in the engine block and require Special Tools for removal and replacement.

OTHER HEAD AND BLOCK ASSEMBLY SERVICING REQUIREMENTS

Except for those areas already mentioned in this section, the G63B head and block assembly closely resemble those of the G52B in design characteristics and servicing requirements.

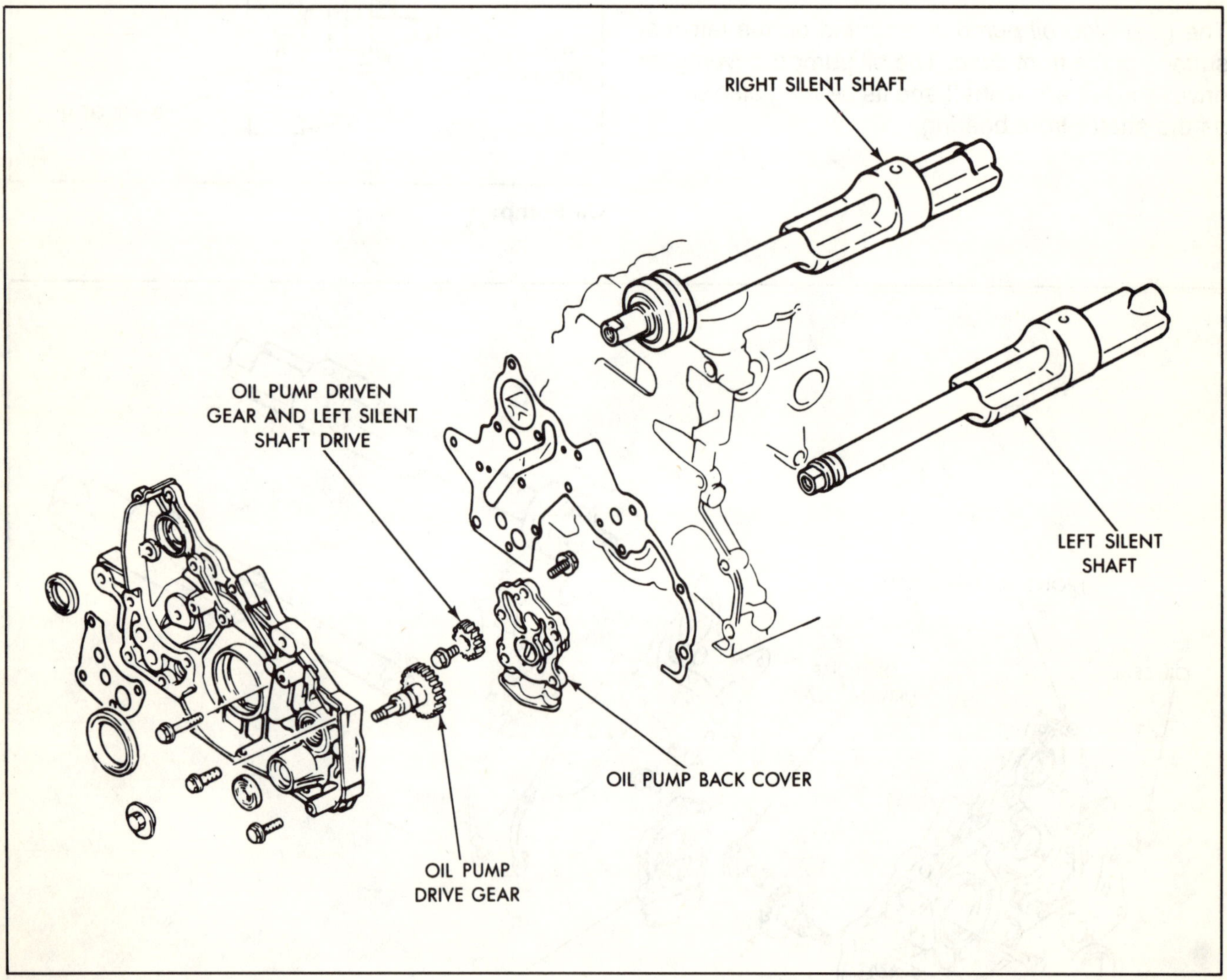

Silent Shaft and Oil Pump

SECTION IV — FUEL SYSTEM

The fuel system of the '83 2.0 Liter engine in external appearance is practically identical to that of the 1982 engine. However it is assumed there might be some later changes in *internal* specifications to meet current emission and driveability requirements. The basic description and operating characteristics remain the same. These are covered, below, for the two major elements of the 1983 fuel system: the *fuel pump* and the *carburetor*.

FUEL PUMP

The fuel pump is a mechanical type mounted on the left side of the cylinder head and driven by a rocker arm which is actuated by an eccentric lobe of the overhead camshaft. In operation, the rocker arm compresses a diaphragm that forms the base of the pump's fuel chamber, forcing fuel through a one-way check valve to the fuel inlet chamber of the carburetor. On the return stroke, a spring draws the diaphragm down, creating a partial vacuum in the fuel chamber that draws fuel from the inlet line through a second one-way check valve, refilling the chamber.

The pump has a second outlet (in addition to that supplying the carburetor's fuel inlet) whose purpose is to purge percolated fuel and fuel vapors from the pump's fuel chamber. This purged fuel is directed to the carburetor's accelerating pump section from where it is carried back to the fuel tank via a return line.

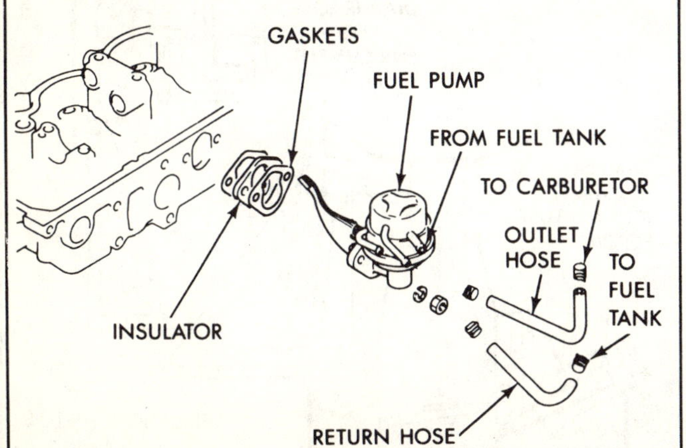

Fuel Pump

The fuel pump's housing is permanently sealed so defective pumps cannot be disassembled for servicing but must be replaced.

CARBURETOR

The '83 engine (like the '82 engine) uses a two-barrel Mikuni carburetor. It includes a *fuel inlet system; primary* and *secondary metering systems,* each with its own venturi and throttle valve; an *enrichment system; accelerator pump;* and *automatic choke systems.*

FUEL INLET SYSTEM

This is a fairly conventional float-in-chamber type with an integral fuel filter at the inlet. It also incorporates a solenoid-operated bowl vent and an anti-overfill device. This latter feature locks the float and inlet needle valve in the closed position to prevent fuel leakage in the event of a roll-over.

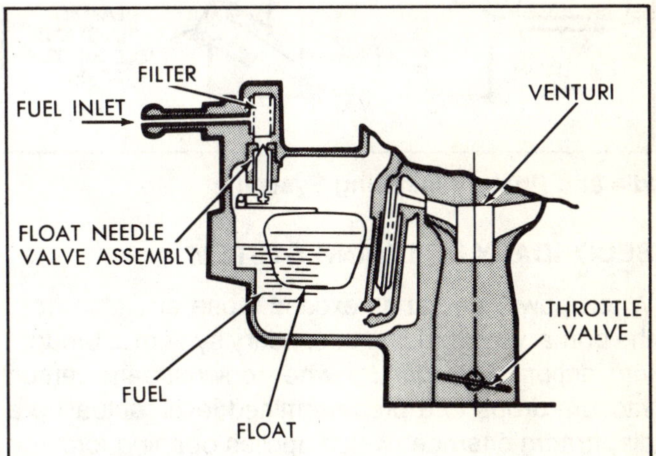

Fuel Inlet System

SECTION IV — FUEL SYSTEM

PRIMARY METERING SYSTEM

This system's venturi and inner venturi nozzle are in operation whenever the engine is running and they are the sole source of fuel-air mixture to the engine up to part-throttle-crusing power settings. The idle and off-idle mixture adjuting screws, shown, are normally pre-set, then sealed to prevent tampering.

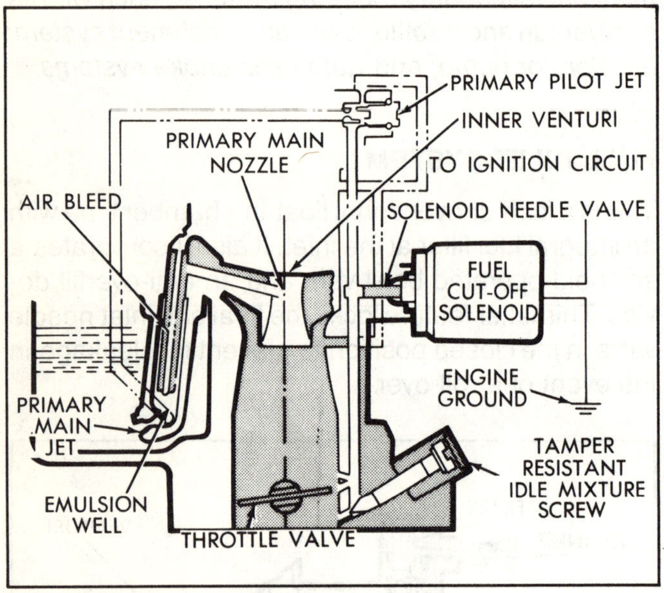

Idle and Primary Metering System

SECONDARY METERING SYSTEM

When power demands exceed those available from the primary system, the secondary system is brought into action. This occurs when downstream venturi vacuum drops to a predetermined level, actuating a diaphragm chamber which applies opening force on the secondary throttle valve. However, an interlock (not shown) delays the secondary opening until the primary throttle is approximately half-way (or more) open.

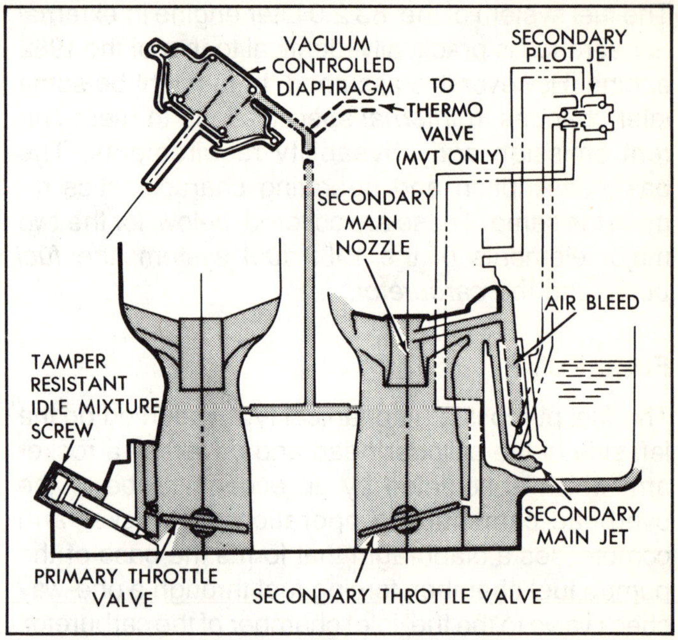

Secondary Metering System

ENRICHMENT SYSTEM

This provides additional fuel to the secondary nozzle for enrichment during high throttle opening operation. It is actuated by a diaphragm which opens an enrichment valve when signaled by a drop in manifold vacuum to levels indicating wide-open or near-wide-open throttle operation.

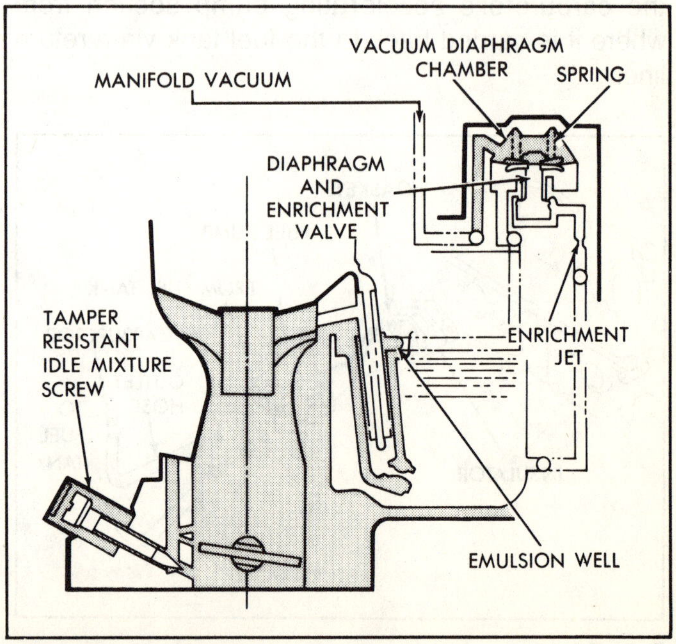

Enrichment System

SECTION IV — FUEL SYSTEM

ACCELERATING PUMP

This consists of a diaphragm-type pump which provides an instantaneous spurt of fuel into the primary system for enrichment needed for prompt power response each time the throttle is opened suddenly.

The follow-up spring, shown, serves to extend the duration of the fuel spurt.

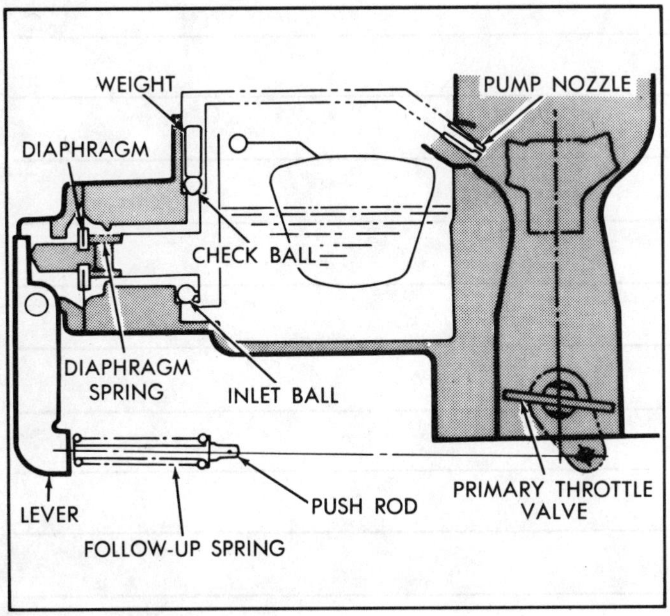

Accelerating Pump System

AUTOMATIC CHOKE

The choke valve is automatically controlled by a thermo-wax element emersed in coolant circulated from the engine. The element increases choke valve opening as coolant temperature rises. To prevent an over-rich mixture during cold starts, the choke valve is partially open by a choke breaker actuated by intake manifold vacuum. A fast idle cam on the choke breaker shaft (not shown) sets the throttle for fast idle during warm-up.

CAUTION: **Before any fuel system disassembly, remove fuel filler cap to release system pressure. Do not attempt any disassembly on hot engine or with battery cables connected. Operating the engine with the air cleaner removed can lean the mixture enough to induce back-firing which could cause injury to face or hands above the carburetor.**

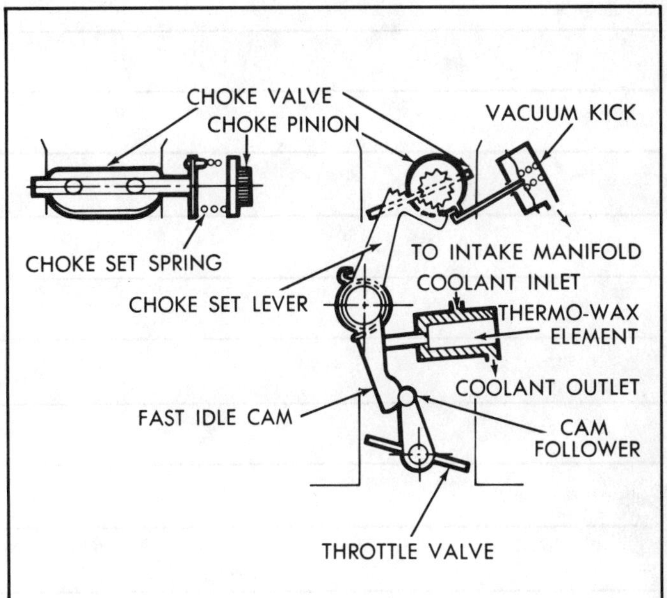

Automatic Choke System

NOTES

SECTION V — EMISSION CONTROL SYSTEMS

The '83 2.0 Liter engine's emission control systems (crankcase, evaporation, exhaust) follow the composition and layout of those used for the '82 engine.

CRANKCASE EMISSION CONTROL SYSTEM

Of conventional design, this system prevents the escape of crankcase vapors to the atmosphere. Its principal components is the PCV valve. At low to moderate power settings, the valve allows fresh air to enter the crankcase and flush out crankcase vapors which are directed to the intake manifold and burned. At high power settings with low intake manifold vacuum (and when the engine is stopped) the valve closes, reversing the flow of crankcase vapors to the air cleaner housing.

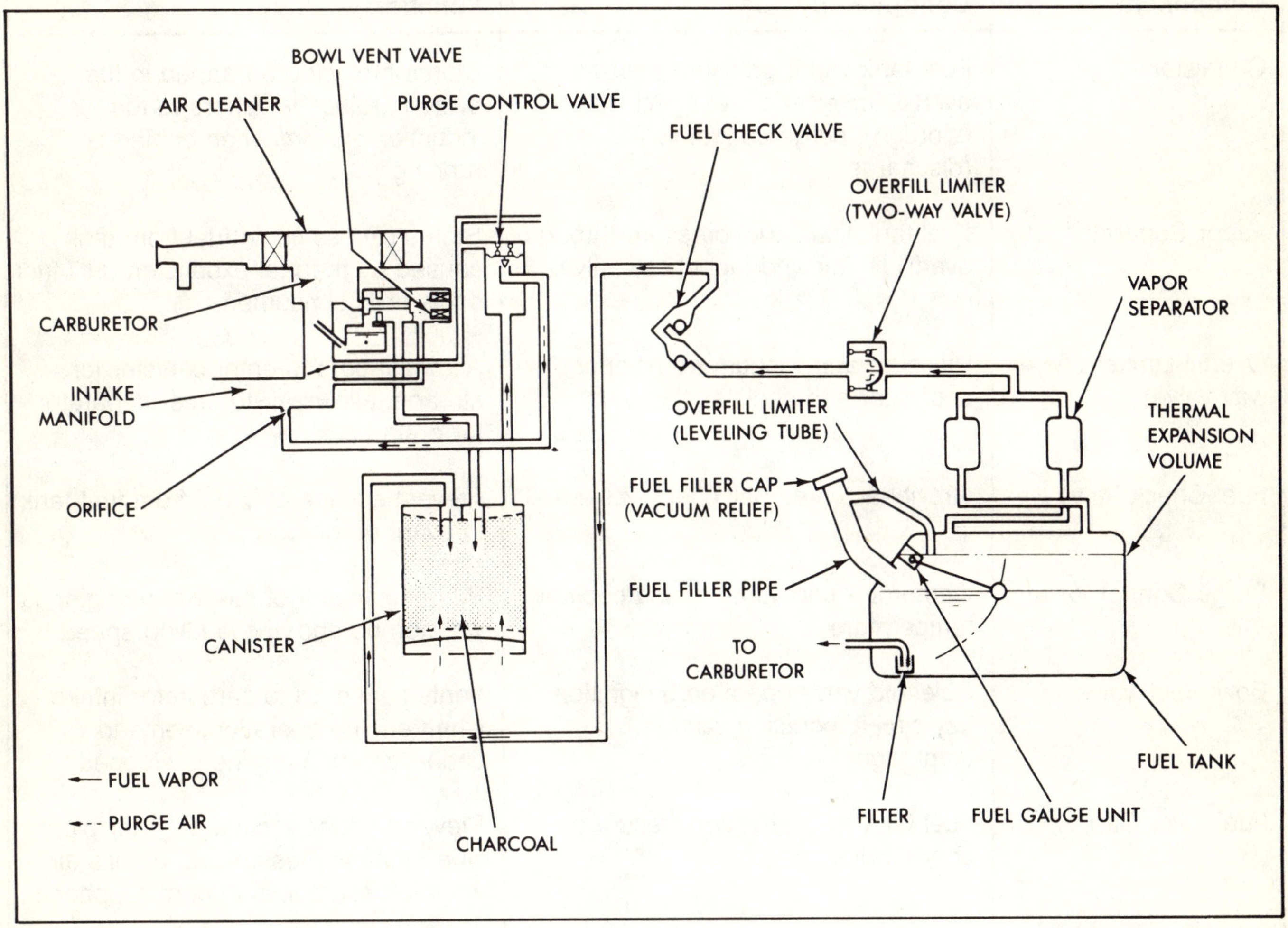

Evaporative Emission Control System — For U.S.A.

SECTION V — EMISSION CONTROL SYSTEMS

EVAPORATIVE EMISSION CONTROL SYSTEM

In effect, this system "seals" the fuel system to prevent the escape of fuel, in either liquid or vapor form, to the atmosphere. Its components are listed below and shown as a system.

EXHAUST EMISSION CONTROL SYSTEM

The exhaust emission control system is designed to improve the *quality* of the exhaust gases by reducing the percentage of unburned *hydrocarbons* (HC) and the percentage of nitrogen oxides (NO_X) in its content. The components of the system can be classified by *which* undesirable emission they help control.

Component	Controlled by	Function
Cannister	Fuel tank vapor pressure through overfill limiter and fuel check valve (storage); purge control valve (discharge)	Stores fuel vapors trapped in fuel system, releases vapors to the induction system when engine is running
Vapor Separator	Fuel tank liquid fuel pressure through overfill limiter and fuel check valve	Stores excess liquid fuel from tank caused by thermal expansion until fuel cools or is consumed
Overfill Limiter (Two-way valve)	Pressure and vacuum from vapor separator and canister	Allows vapors to enter canister for storage, allows air to enter separator for drain back
Fuel Check Valve	Gravity-actuated ball check valves	Prevent escape of liquid from fuel tank in event of roll-over
Purge Control Valve	Carburetor port vacuum and coolant temperature	Purges canister of fuel when engine is warmed up and above idling speed
Bowl Vent Valve	Solenoid valve operated by ignition key circuit induction vacuum diaphragm	Vents float bowl to carburetor intake when engine operation; vents to cannister when engine is stopped
Fuel Filler Cap	Fuel tank air and vapor presure on check valve	Prevents vapor escape out filler pipe when tank is pressurized, admits air when pressure less than atmospheric
Carbon element (in air cleaner element)	Fuel vapors	Absorbs vapors from residual fuel in carb barrels and intake manifold

SECTION V — EMISSION CONTROL SYSTEMS

HC AND CO CONTROL COMPONENTS

HEATED AIR INTAKE SYSTEM

Helps improve fuel atomization during warm up by drawing induction air from around the exhaust manifold to assure better fuel-air mixing for more complete combustion. As the engine warms, a vacuum motor controlled by air manifold vacuum and temperature shuts off heated air to carburetor and allows cooler outside air to be used for better volumetric efficiency.

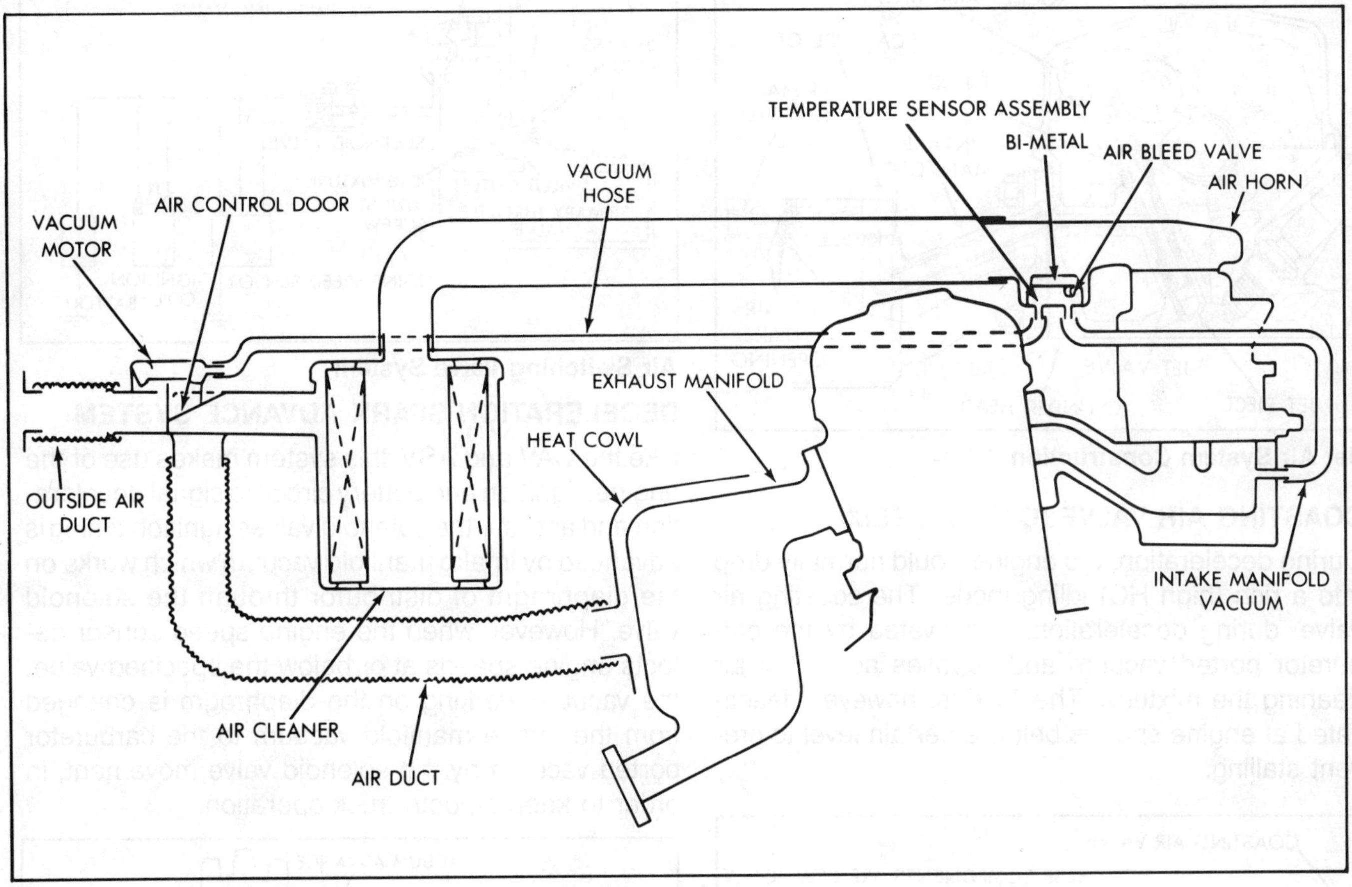

Heated Inlet Air System Hoses and Duct Routing (Typical)

SECTION V — EMISSION CONTROL SYSTEMS

JET VALVES

The valves which are synchronized to open and close with the regular intake valves help insure more complete combustion by inducing a swirling mixture in the combustion chamber that helps complete the fuel-air mixing process and also helps scavenge partially burned fuel residues.

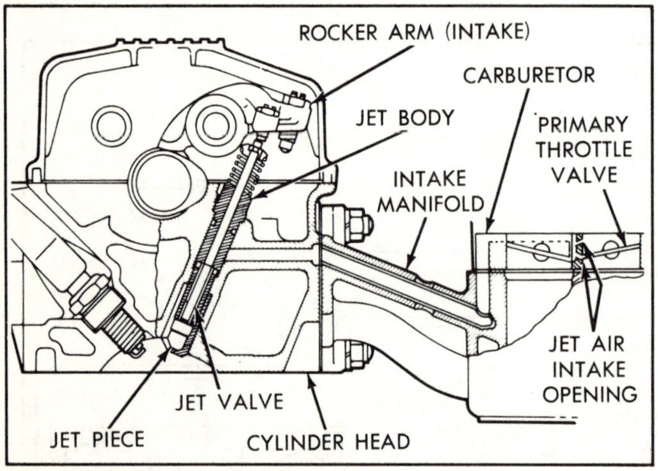

Jet Air System Construction

COASTING AIR VALVE (CAV) SYSTEM

During deceleration, the engine would normally drop into a rich (high HC) idling mode. The coasting air valve, during deceleration, is activated by the carburetor ported vacuum and supplies additional air (leaning the mixture). The CAV is, however, deactivated at engine speeds below a certain level to prevent stalling.

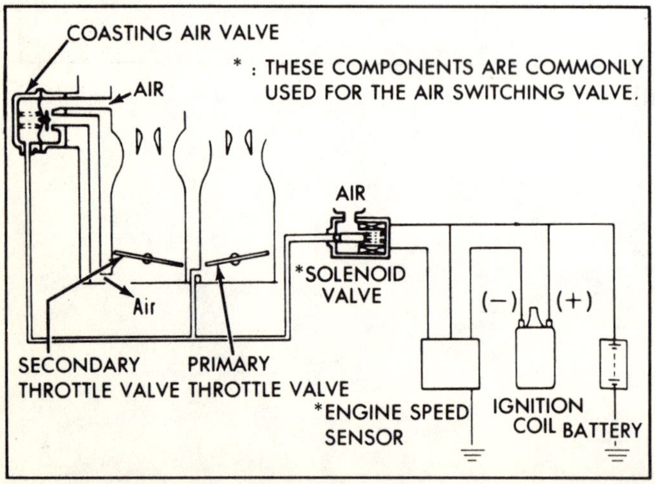

Coasting Air Valve System

AIR SWITCHING VALVE (ASV) SYSTEM

Used in conjunction with the CAV, the ASV limits fuel flow during deceleration to further insure minimum HC formation during deceleration. It, too, is deactivated below certain engine speeds to prevent stalling.

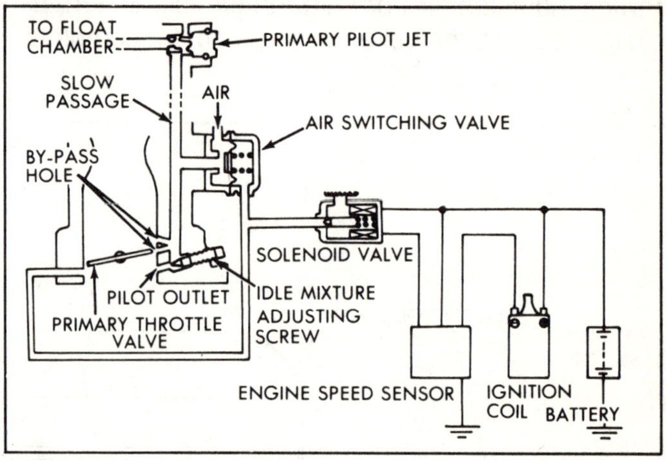

Air Switching Valve System

DECELERATION SPARK ADVANCE SYSTEM

Like the CAV and ASV, this system makes use of the engines' ignition and battery circuit to signal deceleration and activate the solenoid valves. Ignition timing is advanced by intake manifold vacuum which works on the diaphragm of distributor through the solenoid valve. However, when the engine speed sensor detects engine speeds at or below the specified value, the vacuum working on the diaphragm is changed from the intake manifold vacuum to the carburetor ported vacuum by the solenoid valve movement, in order to keep smooth truck operation.

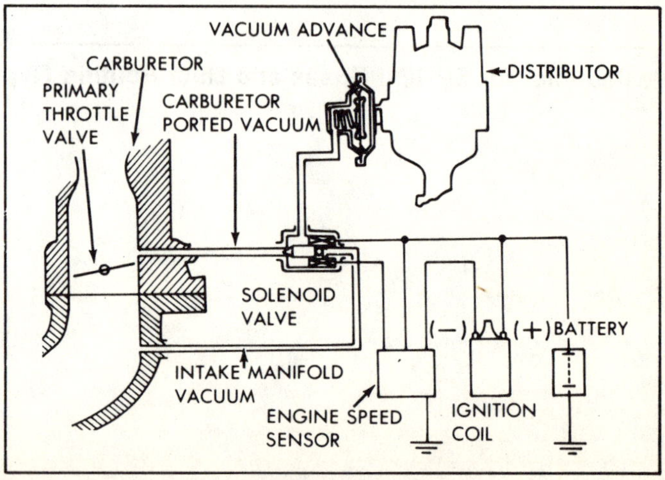

Deceleration Spark Advance System

SECTION V — EMISSION CONTROL SYSTEMS

DASHPOT (On California trucks with manual transmissions)

This device slows the closure rate of the throttle when the throttle is released for deceleration. This helps reduce HC formation during the engine braking mode.

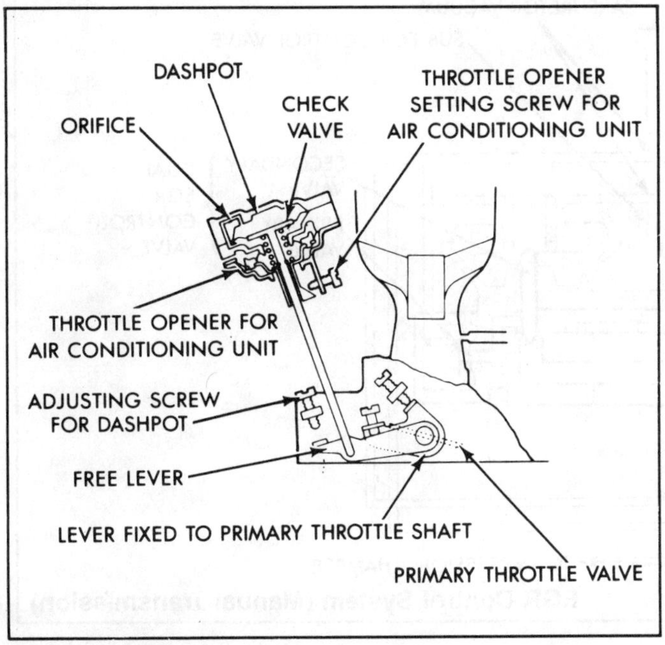

Dashpot Adjustment

CATALYTIC CONVERTER AND SECONDARY AIR SUPPLY SYSTEMS (Dual Systems in California)

The converter is the final component in the system to complete the combustion of HC and CO fuel elements. To provide the additional oxygen needed for burning the HC and CO, a secondary air supply directs air from the air cleaner through a reed-type check valve to the inlet of the converter. The pumping action occurs during the short low-pressure interval following each high-pressure exhaust pulsation, with the check valve action of the reed valve allowing air to move *toward* the converter, only, but preventing exhaust gases from flowing back into the air cleaner.

On California trucks, a dual system is provided with a second catalytic converter and secondary air supply in series below the primary converter.

CAUTION: **Damage to the catalytic converter(s) can result if the truck is not kept in the proper operating conditioning. In the event of an engine malfunction, particularly one involving engine misfire or other apparent loss of performance, have the truck serviced promptly. Continued operation of the truck with a severe malfunction could cause the converter to overheat, resulting in possible damage to the converter and to the truck.**

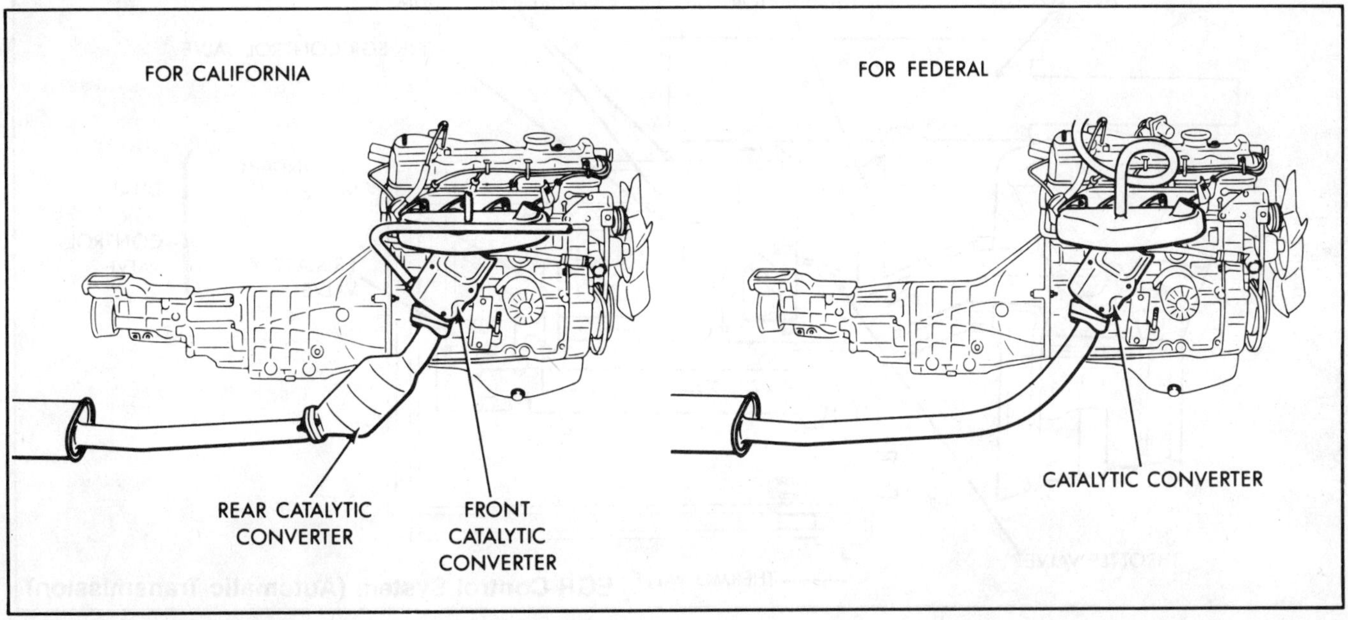

Catalytic Converter

SECTION V — EMISSION CONTROL SYSTEMS

NO_x CONTROL COMPONENTS

The nitrogen in air (N_2) while normally relatively inert, chemically speaking, can unite with the air's oxygen to form various oxides of nitrogen in an engine's combustion chamber if temperatures are high enough. The EGR system, accordingly, acts to reduce combustion chamber temperatures by diluting the intake mixture with a metered amount of exhaust gas which having already been burned helps lower temperatures. The sysems EGR valves control the amount of exhaust gas added to the intake mixture.

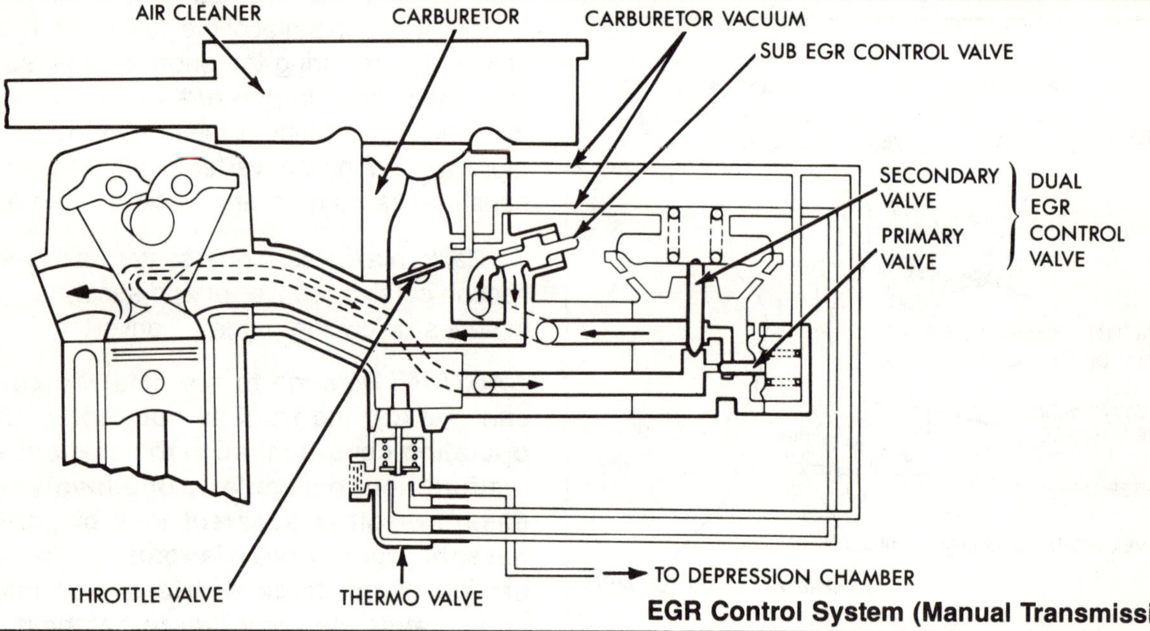

EGR Control System (Manual Transmission)

EGR Federal

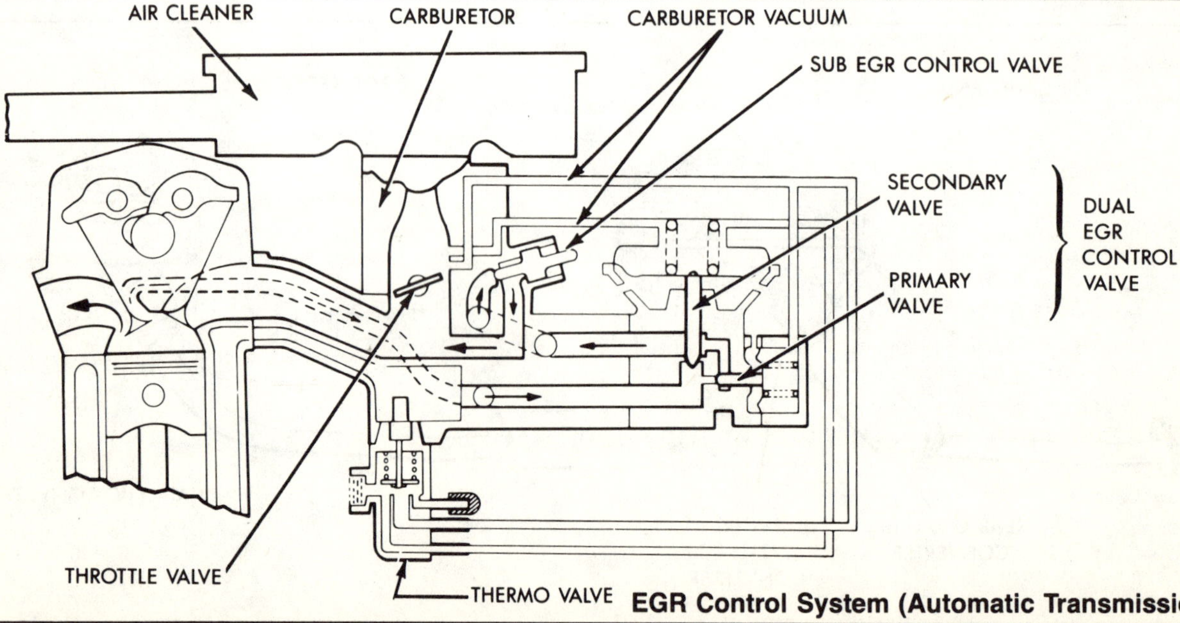

EGR Control System (Automatic Transmission)

EGR 50 States

SECTION V — EMISSION CONTROL SYSTEMS

DUAL EGR CONTROL VALVE (Exhaust Gas Recirculation)

The valve is controlled by two separate carburetor vacuum sources, each of which is moderated by a thermo valve. At low power settings above idle, the primary valve is open to allow exhaust gas into the intake system. At higher settings, a secondary valve will also open allowing a larger flow of exhaust gas.

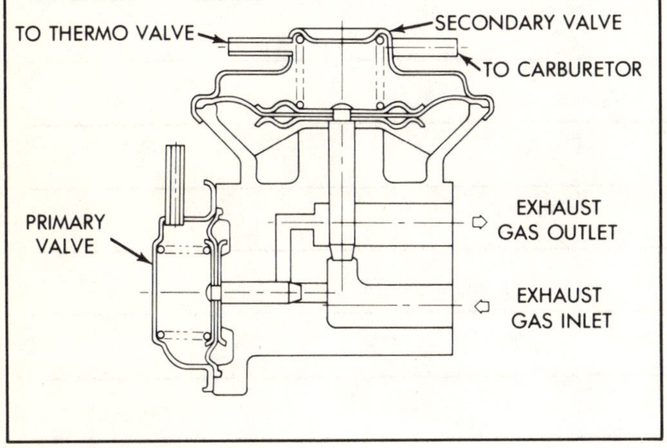

Dual EGR Control Valve

SUB EGR VALVE

This valve is in series with the dual EGR valve and is controlled directly by the throttle valve to provide a throttle opening modulation of the EGR flow to the engine.

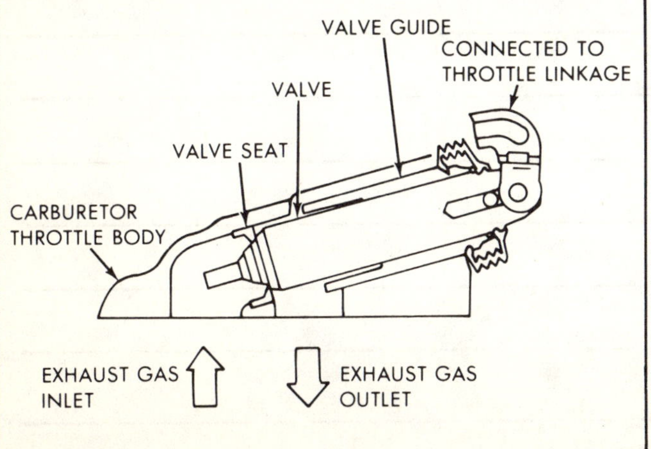

Sub EGR Control Valve

THERMO VALVE

Since, during warm up, cold head temperatures also limit combustion chamber temperatures, there's no need for EGR until the engine reaches its normal operating temperature. This cold-engine cutoff of EGR is handled by the thermo valve which shuts off carb vacuum to the dual EGR valve until the engine coolant reaches operating temperatures.

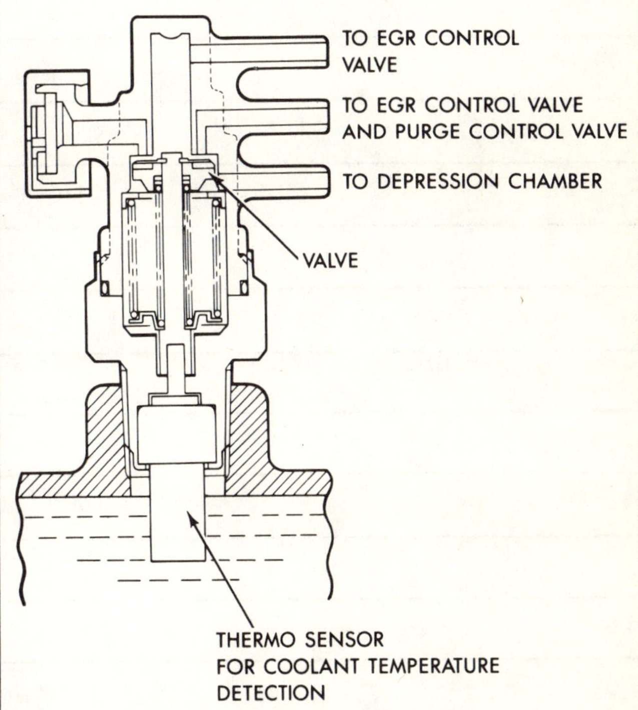

Thermo Valve (Applicable to 2.0 TF with Manual Transmission Vehicles)

NOTES

SECTION VI — IGNITION SYSTEM

The 1983 2.0 Liter engine ignition system is similar to that of the '82 engine employing an ignition coil and a distributor. The distributor is conventional in that it consists of three sections that provide the function common to all distributors:

- A high-voltage distribution section
- A low-voltage breaker section
- An ignition timing advance section

HIGH-VOLTAGE DISTRIBUTION SECTION

Following typical practice, this section uses a rotor turning at half engine speed which directs high-tension pulses to spark plugs lead terminals in the distributor cap.

IGNITION ADVANCE TIMING SECTION

Also conventional in that it uses centrifugal flyweights and on *automatic*-transmission engines, a *single* carburetor port vacuum diaphragm to provide an advance curve based on engine speed and induction air vacuum.

With *manual*-transmission engines a *dual*-diaphragm is used which adds intake manifold vacuum to the advance system during periods of closed throttle deceleration or idling.

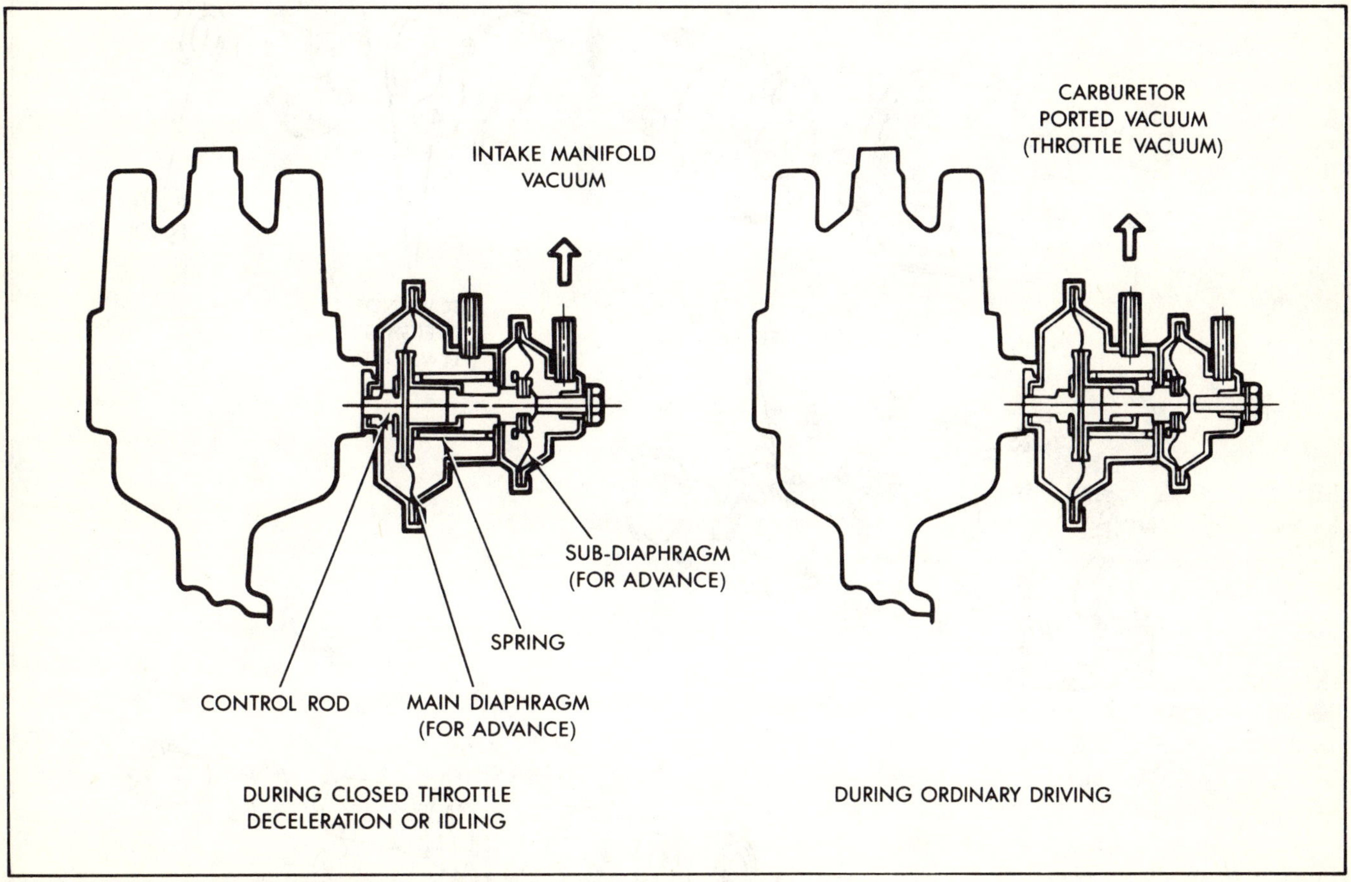

Dual Diaphragm Ignition Advance (Manual Transmission Only)

SECTION VI — IGNITION SYSTEM

LOW-VOLTAGE BREAKER SYSTEM

This system follows modern practice in which the troublesome mechanical breaker points are replaced by a totally-electronic switching circuit. It consists of three elements:

- A *Signal Rotor* which turns at half engine speed and incorporates four equally-spaced magnets which rotate past a pickup coil.
- The *Pickup Coil* when each magnet passes, it sends a timed electronic impulse to the IC Ignitor.
- The IC (Integrated Circuit) Ignitor is a hermetically-sealed electronic which, when triggered by the pickup coil, opens and closes the low-voltage circuit to the primary of the ignition coil.

While the distributor requires little if any adjustment or servicing, if disassembly is necessary, two cautions should be observed:

- Do not remove the grease on the back surface of the IC Ignitor. This grease is necessary to provide adequate heat radiation from the ignitor.
- Make sure there are no metallic dust or particles adhering to the breaker assembly when it is reassembled. These could alter the timing or cause short circuiting.

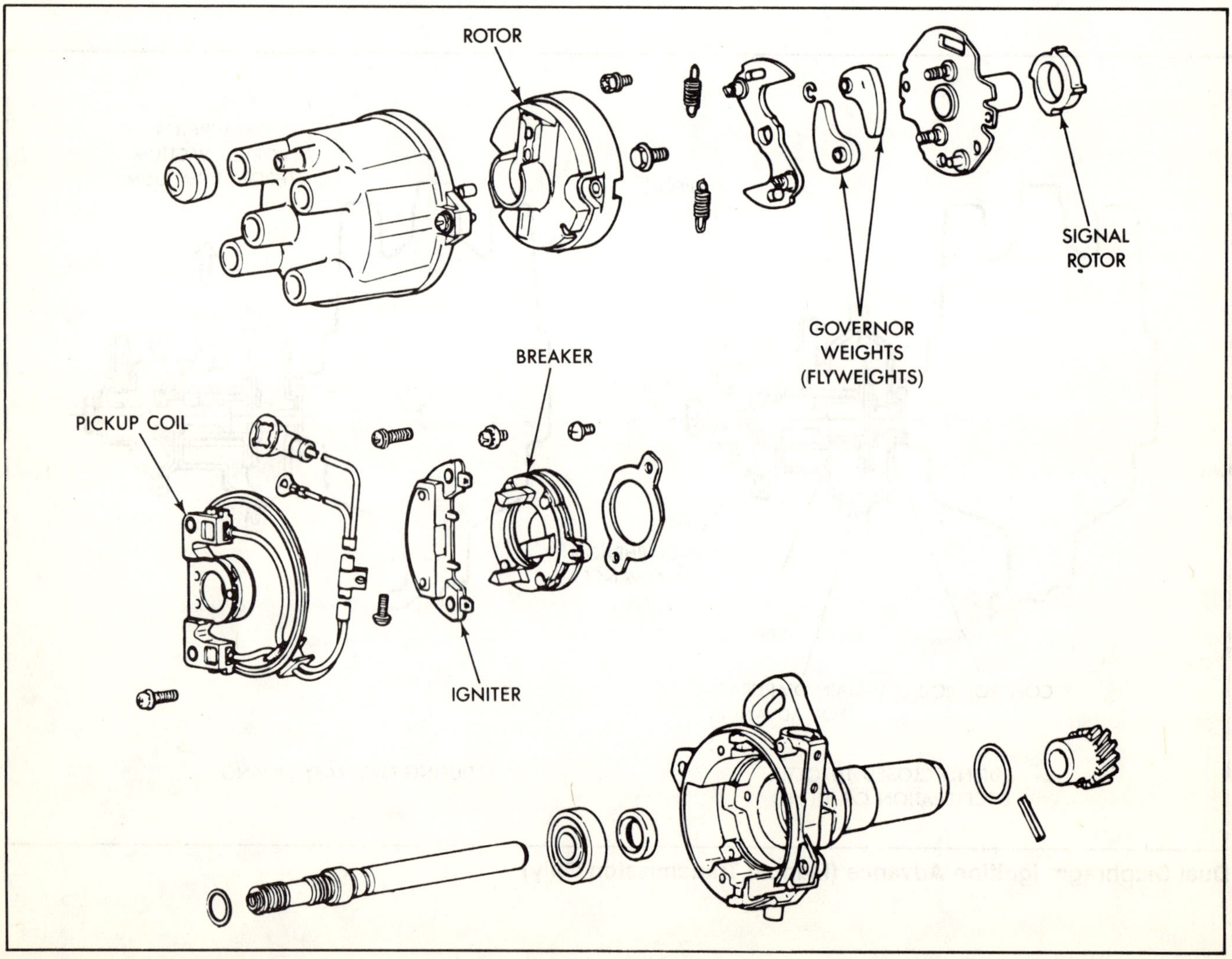

Distributor

SECTION VII — MAJOR TIGHTENING TORQUES AND ADJUSTMENT CLEARANCES

TIGHTENING TORQUES

Cold engine . . . . 89-98 N•m (65-72 ft-lbs)
Hot engine . . . . 98-107 (73-79 ft-lbs)

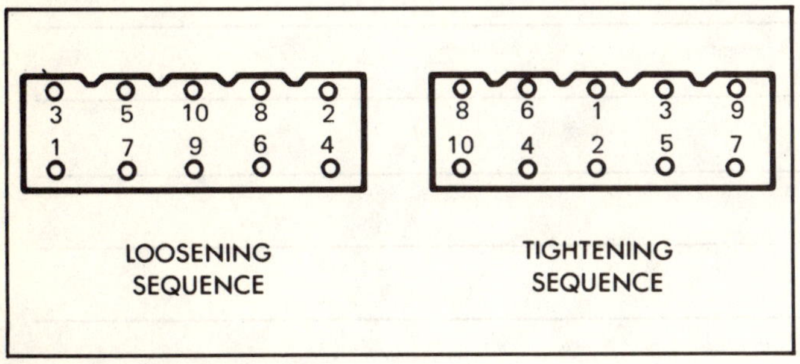

Tightening Torques

INTAKE-EXHAUST MANIFOLD . . . . 15-19 N•m (11-14 ft-lbs)

SPARK PLUG . . . . 20-29 N•m (15-21 ft-lbs)

ROCKER COVER . . . . 5-6 N•m (4-5 ft-lbs)

OIL PUMP SPROCKET . . . . 34-39 N•m (25-28 ft-lbs)

OIL FILTER . . . . 11-12 N•m (8-9 ft-lbs)

VALVE ADJUSTMENT CLEARANCES (Cylinder TDC on compression stroke)

Intake valve . . . . 0.15mm (.006 in.)
Jet valve . . . . 0.15mm (.006 in.)
Exhaust valve . . . . 0.25mm (.010 in.)

BASIC IGNITION TIMING . . . . 5° BTDC

OIL PAN . . . . 6-7 N•m (4.5-5.5 ft-lbs)

OIL PAN DRAIN PLUG . . . . 59-78 N•m (44-57 ft-lbs)

CAMSHAFT BEARING CAP . . . . 17-20 N•m (14-16 ft-lbs)

CAMSHAFT TIMING SPROCKET . . . . 79-98 N•m (59-72 ft-lbs)

JET VALVE ASSEMBLY . . . . 18-21 N•m (13-15 ft-lbs)

NOTES

SECTION VIII — G63B 2.0L ENGINE SPECIFICATIONS

Some of the significant 1983 specifications differing from those of 1982 are indicated with an asterisk (*) and show the comparable 1982 specification to the right in brackets [ ].

Description	1983 Specifications
General	
No. and arrangement of cylinder	4, in-line, vertical
Combustion chamber type	Hemi-spherical
Valve arrangement	Overhead valve type
Camshaft arrangement	Overhead camshaft type
Total displacement cc (cu. in.)	1,997 (121.8) [1,995 (121.7)]
Bore x Stroke mm (in.)	85 x 88 (3.35 x 3.46) [84 x 90 (3.31 x 3.54)]
Compression ratio	8.5
Valve timing	
Inlet valve and jet valve opened/closed	19°BTDC/57°ABDC* [25°BTDC/59°ABDC]
Exhaust valve opened/closed	57°BBDC/19°ATDC* [70°BTDC/14°ABDC]
Firing order	1—3—4—2
Valve clearance at hot engine mm (in.)	0.15 (.006) — Inlet and jet 0.25 (.010) — Exhaust
Timing belt (front)	
Type	Clogged belt type* [Chair type]
No. of teeth	122* [102]
Width mm (in.)	19 (.75)* [9.525]
Timing belt "B" (rear)	
Type	Cogged belt type* [chain type]
No. of teeth	55*
Width mm (in.)	12.7 (.5)*
Crankshaft sprocket	
Material	Ferrous sintered alloy
No. of teeth	24*
Crankshaft sprocket "B"	
Material	Ferrous sintered alloy
No. of teeth	32*
Camshaft sprocket	
Material	Cast iron
No. of teeth	48*
Oil pump sprocket	
Material	Ferrous sintered alloy
No. of teeth	18*
Silent shaft sprocket	
Material	Ferrous sintered alloy
No. of teeth	16*

SECTION VIII — G63B 2.0L ENGINE SPECIFICATIONS

Description	1983 Specifications
Rocker arm	
Material	Aluminum die casting
	Slipper made from special sintered alloy
I.D. mm (in.)	18.9 (.744)
Oil clearance mm (in.)	0.01 to 0.04 (.0004 to .0016)
Camshaft	
Driven by	Cogged type belt* [chain]
Material	Cast iron, Cam surface chilled
Cam height mm (in.)	42.2 (1.661)
Valve lift mm (in.)	9.9 (.390)
Cam diameter for fuel pump drive mm (in.)	38 (1.496)
Journal diameter mm (in.)	34 (1.339)
Oil clearance mm (in.)	0.05 to 0.09 (.002 to .004)
Identification mark	1
Cylinder head	
Material	Aluminum alloy
Flatness of gasket surface mm (in.)	Max. 0.1 (.004)
Valve guide hole diameter mm (in.)	
0.05 (.002) O.S.	13.050 to 13.068 (.5138 to .5145)
0.25 (.010) O.S.	13.250 to 13.268 (.5217 to .5224)
0.50 (.020) O.S.	13.500 to 13.518 (.5315 to .5322)
Inlet valve seat ring hole diameter mm (in.)	
0.3 (.012) O.S.	44.300 to 44.325 (1.7441 to 1.7451)
0.6 (.024) O.S.	44.600 to 44.625 (1.7559 to 1.7569)
Exhaust valve seat ring hole diameter mm (in.)	
0.3 (.012) O.S.	38.300 to 38.325 (1.5079 to 1,5089)
0.6 (.024) O.S.	38.600 to 38.625 (1.5197 to 1.5206)
Valve guide installing height mm (in.)	15 (.591)
Inlet valve	
Material	Special heat resisting steel
Treatment	Tufftriding
Valve diameter mm (in.)	43 (1.693)
Stem diameter mm (in.)	8.0 (.3150)
Clearance (Stem-to-guide) mm (in.)	0.03 to 0.06 (.0012 to .0024)
Margin mm (in.)	1.2 (.047)
Identification mark	6

SECTION VIII — G63B 2.0L ENGINE SPECIFICATIONS

Description	1983 Specifications
Exhaust valve	
Material	Special heat resisting steel Valve face padded with stellite
Treatment	Tufftriding
Valve diameter mm (in.)	35 (1.378)
Stem diameter mm (in.)	8.0 (.3150)
Clearance (Stem-to-guide) mm (in.)	0.05 to 0.09 (.0020 to .0035)
Margin mm (in.)	2.0 (.079)
Identification mark	5S
Valve spring	
Free height mm (in.)	47.5 (1.870)
Load N(lbs.)/mm (in.)	275 (62)/40.4 (1.591)
Square	Less than 1.5°
Identification color	Blue
Right silent shaft	
Driven by	Cogged belt* [Oil Pump Driven Sprocket]
Material	Steel
Front journal diameter mm (in.)	42 (1.654)
Rear journal diameter mm (in.)	41 (1.614)
Oil Clearance mm (in.)	
Front	0.02 to 0.06 (.0008 to .0024)
Rear	0.05 to 0.09 (.0020 to .0035)
Left silent shaft	
Driven by	Oil Pump Driven Sprocket* [Chain]
Material	Steel
Front journal diameter mm (in.)	18.5 (.728)
Rear journal diameter mm (in.)	41 (1.614)
Oil clearance mm (in.)	
Front	0.02 to 0.05 (.0008 to .0020)
Rear	0.05 to 0.09 (.0020 to .0035)
Piston	
Material	Special aluminum alloy
Type	Autothermic (Steel strut incorporated)
Clearance (Piston-to-cylinder) mm (in.)	0.02 to 0.04 (.0008 to .0016)
Pistons for service mm (in.)	0.25 (.010), 0.50 (.020), 0.75 (.030), 1.00 (.040), oversize

SECTION VIII — G63B 2.0L ENGINE SPECIFICATIONS

Description	1983 Specifications
Piston ring	
No. of rings per piston	
Compression	2
Oil	1
Compression ring type	
No. 1	Barrel type, special cast iron, chrome face
No. 2	Taper type, special cast iron
Oil ring type	3-piece steel rail chrome face
Ring gap mm (in.)	
No. 1	0.25 to 0.45 (.010 to .018)
No. 2	0.2 to 0.4 (.008 to .016)
Oil	0.2 to 0.5 (.008 to .020)
Ring side clearance mm (in.)	
No. 1	0.05 to 0.09 (.002 to .004)
No. 2	0.02 to 0.06 (.001 to .002)
Rings for service mm (in.)	0.25 (.010), 0.50 (.020), 0.75 (.030), 1.00 (.040), oversize
Connecting rod	
Length (Center to center) mm (in.)	150 (5.906)
Piston pin bore diameter mm (in.)	20.974 to 20.985 (.8257 to .8262)
Side clearance (Big end) mm (in.)	0.1 to 0.25 (.004 to .010)
Crankshaft	
Material	Forged Steel
Counterweights	8* [.4]
Main bearing journal diameter mm (in.)	57 (2.244)* [60, 2.362]
Connecting rod journal diamter mm (in.)	45 (1.772)* [53, 2.087]
Maximum allowable out-of-round and/or taper of journal mm (in.)	0.01 (.0004)
Oil clearance mm (in.)	
Main bearing journal	0.02 to 0.05 (.0008 to .0020)
Connecting rod journal	0.02 to 0.05 (.0008 to .0020)
Thrust taken by	No. 3 main bearing
End play mm (in.)	0.05 to 0.18 (.0020 to .0071)
Bearings for service available in standard size and following undersize mm (in.)	0.25 (.010), 0.50 (.020), 0.75 (.030)

SECTION VIII — G63B 2.0L ENGINE SPECIFICATIONS

Description	1983 Specifications
Cylinder Block	
Material	Cast iron
Cylinders	Siamese type [Separate]
Out-of-round and taper mm (in.)	Less than 0.02 (.0008)
Maximum allowable oversize (Cylinder bore) mm (in.)	1.00 (.039)
Oil pump	
Type	Gear
Driven by	Cogged type belt [Chain]
Relief valve opening pressure kPa (psi)	440 (64)
Oil filter	
Type	Cartridge, Full flow
Size (Diameter x Length) mm (in.)	90 x 100 (3.54 x 3.94)
Engine oil	
Capacity including that of oil filter lit. (U.S. qts., Imp. qts.)	
Rear-wheel drive models	4.0 (4.2, 3.5)
4-wheel drive models	5.0 (5.2, 4.4)
Recommended oil (API classification)	SE or SF

NOTES

NOTES

NOTES